Cassava - Recent Updates on Food, Feed, and Industry

Edited by Andri Frediansyah

Published in London, United Kingdom

Cassava - Recent Updates on Food, Feed, and Industry
http://dx.doi.org/10.5772/intechopen.105238
Edited by Andri Frediansyah

Contributors
Andri Frediansyah, Ardiba Rakhmi Sefrienda, Dedy Kurnianto, Jasmadi Jasmadi, Tagesse Tadesse, Amanze Ngozi Joan, Abah Simon Peter, Bariot Hafif, Yulia Pujiharti, Alvi Yani, Noveria Sjafrina, Robet Asnawi, Nendyo Adhi Wibowo, Neneng Laela Nurida, Ai Dariah, Ruth Naa Ashiokai Prempeh, Victor Acheampong Amankwaah, Allen Oppong, Marian Dorcas Quain, Emmanuel Osewe Akubor, Beatrice Amili Akubor, Evaristus Funmilayo Awoyera, Kariuki Samwel Muiruri, Anwar Aliya Anwar, Benson Oloya, Christopher Adaku, Morgan Andama, Stephen Kwame Torkpo, Emmanuel Amponsah, Tristam Pascal Moeliono, Koerniatmanto Soetoprawiro, Wen Feng, Hai-Tian Fu, Yan-Chun Luo, Jian-Q Huang

First published in London, United Kingdom, 2024 by IntechOpen
IntechOpen is the global imprint of INTECHOPEN LIMITED, registered in England and Wales, registration number: 11086078, 5 Princes Gate Court, London, SW7 2QJ, United Kingdom

British Library Cataloguing-in-Publication Data
A catalogue record for this book is available from the British Library

Additional hard and PDF copies can be obtained from orders@intechopen.com

Cassava - Recent Updates on Food, Feed, and Industry
Edited by Andri Frediansyah
p. cm.
Print ISBN 978-1-83768-703-9
Online ISBN 978-1-83768-704-6
eBook (PDF) ISBN 978-1-83768-705-3

For EU product safety concerns:
IN TECH d.o.o., Prolaz Marije Krucifikse Kozulić 3, 51000 Rijeka, Croatia,
info@intechopen.com or visit our website at intechopen.com.

Meet the editor

Dr. Andri Frediansyah is a senior researcher at the Research Center for Food Technology and Processing (PRTPP), Research Organization of Food and Agriculture (ORPP), which is part of the National Research and Innovation Agency (BRIN), Republic of Indonesia. His areas of interest in research range from the fundamentals to the applied sciences of natural products and substances that are related to food and human health. He is also an active member of the Indonesian Researcher's Union *(Perhimpunan Periset Indonesia)*. Dr. Frediansyah graduated with a BSc in Biology from Universitas Gadjah Mada (UGM), Indonesia in 2010. In 2015, he earned his MSc in Applied Biological Sciences from Chulabhorn Graduate Institute, Thailand. In 2020, he obtained a Ph.D. (magna cum laude) in Pharmaceutical Biology from the Eberhard Karl University of Tübingen, Germany.In addition, he is an experienced writer who has contributed to numerous scientific publications. He is also a recipient of numerous awards from the scientific community, including the 2023 J. William Fulbright Visiting Scholar Award.

Contents

Preface

Cassava is a woody perennial shrub that grows in tropical and subtropical regions of the globe and has an edible root. Cassava originated in tropical America, was first introduced to sub-Saharan Africa, and subsequently spread to other regions of the world, including tropical Asia. It contains countless carbohydrates, vitamins, and minerals. However, nutrient composition varies depending on the age and type of the harvested crop, as well as climate, soil conditions, and other environmental factors present during cultivation. In light of these facts, cassava is a significant crop in terms of providing energy and nutrition as human food or animal nourishment as well as a resource for multiple industries. Additionally, cassava is significant for food security and the circular economy in several countries around the world. Therefore, there is no end to researching and learning about the existence of cassava from various perspectives.

This book focuses on (1) the global perspective of cassava, food security, and the circular economy; (2) biotechnology and soil management; (3) disease control; and (4) the processing of cassava in food, feed, and multiple industries. These are several updates that determine the future of cassava crops and their worth as agricultural commodities in various nations. Without a circular economy and perspective, cassava will cease to exist in many countries as a staple nutrition. Technological intervention updates, including plant biotechnology, disease control, and soil management, are required to ensure sufficient cassava production to meet consumer demand for high-quality products. Lastly, the development of novel foods, feed products, and multiple industries could increase the market value of cassava products. The most critical challenge in the post-harvest cassava industry is the development of new products as part of the diversification process.

I would like to express my sincere appreciation to the authors who contributed so generously to this book; without them, this endeavor would not have been successful. In addition, I wish to express my appreciation to IntechOpen, with whom I have collaborated on similar projects. This has always been a pleasant experience, and I look forward to collaborating with them in the future on many more initiatives. Finally, I thank Publishing Process Managers Ms. Nika Karamatic and Ana Javor for their invaluable assistance in assembling the materials.

In conclusion, we hope that both scientific and non-scientific communities will find this book helpful for making informed decisions about cassava and perceiving it as a crop that would be advantageous to producing novel foods, feeds, and multiple industries.

Andri Frediansyah
Research Center for Food Technology and Processing (PRTPP),
National Research and Innovation Agency (BRIN),
Yogyakarta, Indonesia

Section 1

Cassava on Global Perspective Food Security and Circular Economy

Chapter 1

Is Cassava the Next Magic Solution to World Hunger and Energy Crisis? The Indonesian Experience

Tristam Pascal Moeliono and Koerniatmanto Soetoprawiro

Abstract

This chapter addresses, from a legal (state law) perspective, the issue regarding the feasibility of using cassava as alternate staple food beside rice for the Indonesian people or, in the context of energy crisis, as source for biofuel. To do that, written formal law touching upon sustainable development, agriculture, food sovereignty-security and other relevant law and regulations will be perused and analyzed. Focus shall be given on the symbolic meaning of written laws and how government officials interpret and further develop it into policies. The main argument here is that while there are local-regional initiatives and even start up policies to develop cassava as flex crop, lessening Indonesian people dependence on rice as staple food, or magic solution to world hunger or as raw material for biofuel, those steps have not yet become part of the formal law and possessing symbolic effect.

Keywords: cassava, food security-sovereignty, staple food, Indonesia, mass hunger

1. Introduction

All countries, including Indonesia, being aware that they cannot avoid dealing with climate change, should also take into consideration the (direct-indirect) consequence of this global phenomenon: such as food crisis, mass hunger, rising sea levels, and all other socioeconomic problems that arise along with it. All these issues have mentioned as part of the SDGs and discussed extensively at the G-10 in Bali (2022). These global problems become worse in the face of armed conflicts (internal-international) occurring in many parts of the world, lastly the open war between Ukraine and Russia, February 24, 2022. The last named successfully disrupt the global production chain of instant noodles and other food products made from wheat flour.

Regardless, mass nationwide food crisis, whatever the cause, has the potential and in fact had led to the downfall of successive government regimes. In fact, Indonesia has experienced widespread nation scale food crises several times and each time it leads to political-social unrest and even the ruling governments downfall. In the 1960s, the regimes bad economic policies leading in fact to food insecurity and widespread hunger mostly in rural areas trigger social unrest and become the driving motivation behind the Three Demands of the People (*Tri Tuntutan Rakyat*).[1] Among other, in addition to the attempted and failed coup in 1965, it causes the Soekarno's government loss of legitimacy. The New Order regime, replacing the Soekarno's regime, prided itself to being able to reach and sustain rice self-sufficiency in the early 1980, falls also due to the impact if nationwide economic crises—causing food insecurity and lack of affordability—ending in a political crisis in the 1997–1998. Apparently providing and securing food becomes one of the main responsibilities of any government wishing to retain political power and legitimacy. Failure to do so, maintaining food security, which sometimes is understood to be interchangeable with food sovereignty, will affect the legitimacy of government to rule.

This chapter explores the Indonesian state cq. government response to the real possibility of food crises. The authors will search, trace, and look at policies, programs, and laws (providing the legal basis to such policies and programs) pertaining to the promotion of cassava as a flex crop, an alternative food source to rice—the Indonesian people supposedly main staple food—and which may be processed into biofuel, expected to help the general population decrease its dependence to unrenewable carbon-based energy sources.

For starters, the crop, cassava, has been known and used as alternative food source or supplement to rice and corn throughout Indonesia and has also been promoted as a flex crop, meeting various human needs outside being a food source. Based on general observation alone, cassava's shoots and tubers in processed form were and still are utilized by people in various regions as side dishes or street food, especially on the island of Sumatera, Java-Bali, and other outer islands, among others the famous Bubur Manado (tinutuan) from North Sulawesi. Their shoots have also been processed as modified cassava flour (mocaf). The waste may albeit in an experimental or non-industrial scale has been processed into biofuel.

2. Research method

The research method used here to tackle the issue raised is a legal perspective or juridical approach. Meaning that formal legal sources (laws, government regulations, ministerial regulations, in short state law) as text will be analyzed. For matters related to Indonesian agriculture and food provision, a systematic analysis of the legal framework had been performed by Koerniatmanto [1] and his legal scholar approach will serve as the foundation to look at the general legal framework related to agriculture and from reading the legal text attempt to discern the changing (if any) policies related to food sovereignty and security, impacting the choice of prioritizing certain crops over the other. On the other hand, any written-formal legal text, albeit interesting from a lawyer viewpoint, may not reveal much. To compensate, legal text in the Indonesian context can and should also be viewed in its wider context. A socio-legal approach will be used here as well. Borrowing from García-Villegas approach, state

[1] Three demands of the Indonesian people comprising of (1) disband the communist party, (2) clean the cabinet from ministers affiliated to the communist party, and lastly (3): lower food and other consumption good prices (1966). The Tritura declaration was drafted on the initiative of students with the support of the military, intending to eradicate communism from Indonesia.

law (and all other formal source of law) is to be understood as a political battle for the right meaning of legal text[2]:

Taking into consideration the above issue, specifically the author peruses and discusses existing laws, regulation or policy statements, or public statements of government officials serving as policy guidelines in the food sector and look at how and when cassava is being considered in terms of food sovereignty or food security or the stated government goal of providing and securing the availability and adequacy of nine basic commodities. This approach may allow us to take a quick look at how the government think and speak of the possibility and feasibility of developing cassava as a flex crop and what programs, policies, or government action plans result from it. The perspective used will also borrow from Tania Li's insight [4] when she traces the persistence of the state's will to secure societal prosperity through various development programs including those that are internationally sponsored. Her approach, on the other hand, seems to disregard the existing legal framework, the official language of the government and bureaucracy. To compensate, the author also borrows from Jaqueline Vel's works [5–7] addressing the issue how development ideas behind any policy or programs intended to increase society's welfare can turn into commodities in the context of global production chain.

3. Paper outline

The first part of this chapter takes a quick look at the findings regarding cassava's potential as staple food (substituting rice) or as an alternative food source. Relevant policies made by international bodies regarding food sovereignty or security or sustainable development will be traced to the extent these have been adopted and translated into national law regulations at the national–local level. In particular, attention will be given on how the provincial government of West Java responds to the call to hail cassava as flex crop, solution to end hunger, or rising energy demand in Indonesia. Here fact shows that this province for long serving as the national rice basket (now being renamed food estates) experienced rising rate of wet paddy field conversion to other uses. The conclusion ends with a summary about how the presence or absence of national-provincial government policies and regulations or at least agriculture improvement programs relates to the success or failure of marketing cassava as the new hype.

4. Cassava as prime commodity or raw material for biofuel?

Most urban-rural people in Indonesia, experienced cassava-tapioca-cassava plant [*Manihot Esculenta*], especially its tubers, as processed food, served as a snack or even street food. The leaves, cooked in various ways, are used as salad or mixed with other ingredients served as side dish. The rest, non-edible parts, or waste product can be processed for animal (sheep-goats) fooder.

[2] García-Villegas [2]. A different notion of symbolic effect of the law is put forward by Bart van Klink. He argues that: (…) the notion of symbolic effets acquires a rather broad meaning here: it covers all the effect that can be traced back to the understanding of one or more legal norms—thus not only effects on a symbolic level (the legal terminology is disseminated, attitudes are changed, awareness of non-legal behavior is increased, etc.) but also those on the "real" level of action (norms addresses may act differently after understanding the law) [p. 137] in Bart van Klink [3].

In some areas, as an exception, processed cassava root is used as a staple food. It was recorded, for example, that people in Cirendeu, a small village in a remote part of West Java, during the colonial period, opted to refuse eating rice as staple food and chose instead to rely on cassava. This was done as a social movement protesting the Dutch colonial government policy of making rice the staple food of the indigenous population. Producing and consuming cassava become a symbol of resistance. During the Japanese occupation (1942–1945), the occupying army mostly confiscated the rice harvest for their own consumption and left village people around central Java and Yogyakarta—surroundings without enough food to survive. The indigenous population were forced to turn to cassava and innovate. What they did, in the end, was mixing cassava with rice, a combination, named Tiwul.

The result was that, even today, food made from cassava products, at least in central-east Java, is considered only for poor people. The rich or well to do prefer rice and show their higher social-economic status that way. Cassava was thus not considered to be on par with rice, a symbol of the well to do segment of the population. This said taken into consideration the exception of Cirendeu people mentioned above and the fact that in the formal-informal market's in a number of regions in Indonesia, cassava products (raw or processed) can be found being offered for sale but again, not as staple food.

Notable is also that, during the Dutch Indies period, Japanese *inter regnum* (1942–1945), and post-independence (1945 to present), cassava was never cultivated on a large scale and only incidentally as a commercial crop. In general, speaking only for Java-Madura-Bali, cassava plantations covering large area are never to be found. That is in contrast to wet/dry rice or paddy fields surrounding small hamlets. Cassava is up to present planted on small plot of land not suitable for rice fields or deliberately planted as hedges or in small plots of land behind houses in rural areas. One big issue, endangering food security (at least sustainable rice production) in Indonesia now is the high rate of conversion of rice fields to other uses. Data show that (**Table 1**)[3]:

With some rare exception in certain provinces (as the number shows above for Papua, West Papua, North Mollucas, South and South East Sulawesi, South Kalimantan, East Nusa Tenggara, Bangka-Belitung and Lampung, which must be compensated with loss of forest land), the alarming trend is the decrease of wet rice (paddy fields), which may lead to a disruption in the rice food supply chain or national food (rice) sovereignty-security. The trend does not stop at 2015. In 2020, the Minister of Agriculture acknowledges the continuing annual decrease of wet rice field throughout Indonesia.[4] Reportedly Ministry of Agriculture, Dr. Ir. Suswono, MMA, stated that up to 2025, the demand for food estates may reach 13.17 million ha. (...) From existing agricultural land (70 million ha), only 45 million is efficiently used. (...) rice-paddy fields are declining with the rate of 50/70 thousand hectares-annually as compared to the making of new rice fields, estimated to 20–40 thousand hectares-annually.

Facing with that threat, the government responded by issuing Presidential Regulation 59/2019 re. control of land use conversion.[5] This regulation is the legal basis for the establishment Tim Pelaksana Pengendalian Alih Fungsi Lahan Sawah

[3] Rate of agricultural land conversion; The Central Statistic Bureau official website, https://www.bps.go.id/indicator/53/179/1/luas-lahan-sawah.html. Data for 2014 is temporary estimation.

[4] Anonimus, Kementan akui lahan sawah berkurang 650 ha ribu ha per tahun, 17 januari 2020 https://www.antaranews.com/berita/1254488/kementan-akui-lahan-sawah-berkurang-650-ribu-ha-per-tahun.

[5] Presidential Regulation 59/2019 re. control of land use conversion.

Province	Rice fields (hectares)		
	2013	2014	2015
Aceh	300,808.00	294,129.00	290337.00
North Sumatera	438,346.00	433,043.00	423,465.00
West Sumatera	224,182.00	225,890.00	226,377.00
Riau	93,338.00	87,594.00	71,910.00
Jambi	113,546.00	101,195.00	94,735.00
South Sumatera	612,424.00	616,753.00	620,632.00
Bengkulu	93,382.00	88,756.00	85,131.00
Lampung	360,237.00	363,055.00	377,463.00
Bangka-Belitung	5358.00	7490.00	10,654.00
Riau Islands	487.00	405.00	246.00
Jakarta	895.00	778.00	650.00
West Java	925,042.00	924,307.00	912,794.00
Central Java	952,980.00	966,647.00	965,262.00
Yogyakarta	55,126.00	54,417.00	53,553.00
East Java	1,102,921.00	1,101,765.00	1,091,752.00
Banten	194,716.00	200,480.00	199,492.00
Bali	78,425.00	76,655.00	75,922.00
West Nusa Tenggara	253,208.00	254,298.00	264,666.00
East Nusa Tenggara	169,063.00	172,954.00	177,238.00
West Kalimantan	330,883.00	323,959.00	330,724.00
Central Kalimantan	225,836.00	215,545.00	196,553.00
South Kalimantan	440,429.00	431,437.00	450,152.00
East Kalimantan	63,323.00	55,485.00	57,000.00
North Kalimantan	21,762.00	21,775.00	21,448.00
North Sulawesi	56,157.00	60,475.00	52,820.00
Central Sulawesi	146,721.00	141,448.00	128,323.00
South Sulawesi	602,728.00	623,139.00	628,148.00
South East Sulawesi	95,378.00	96,826.00	103,812.00
Gorontalo	32,239.00	32,116.00	32,058.00
West Sulawesi	61,070.00	62,312.00	61,292.00
Mollucas	15,042.00	13,519.00	13,394.00
North Mollucas	10,510.00	10,516.00	11,802.00
West Papua	9587.00	9587.00	10,126.00
Papua	42,350.00	42,843.00	44,462.00
Indonesia	8,128,499.00	8,111,593.00	8,087,393.00

Table 1.
Rice field conversion per province.

(ad hoc team authorized to control the wet rice field conversion) headed by the Director General of Land and Spatial Control and Order at the Ministry of Agrarian Affairs and Spatial Planning/National Land Agency (ATR/BPN; Direktur Jenderal Pengendalian dan Penertiban Tanah dan Ruang Kementerian Agraria dan Tata Ruang/ Badan Pertanahan Nasional). Nonetheless, all this effort seems to be in vain as the government fails to employ the monetary incentive-disincentive mechanism[6] to induce regional governments or farm-land owners not to convert productive wet-rice fields. On the other hand, the regional government, specifically those who want to industrialize, are more prone to side with investors and allow productive farm land to be converted into industrial estates.

In addition, the same government also enact and enforce Law 11/2020 (job creation law, replaced by Government Regulation in Lieu of Law no. 2/2002) containing a ruling enabling the government to convert farm or agricultural land (including those productive wet-rice paddy fields found in the so-called food estates (*lumbung pangan*)) of Indonesia in the name of public interest and the realization of national strategic development projects (*pembangunan proyek strategis nasional*). The Law 11/2020 added these new criteria for land acquisition in the public interest as found listed in Law 2/2012 re. land acquisition for development in the public interest.[7] Here, two different policies, protection of productive agricultural land and support more economically and profitable land use, seem to be conflicting.

In any case, spatial and land allocation for cassava was and is never an issue, considering the relative unimportance of the crop in terms of food sovereignty or security. In comparison, by looking at existing government policies, rules and regulations regarding the procurement, supply and maintenance of reserves for the so-called 9 basic commodities, similar remarks may be made. These are rice, sago, and corn; sugar; vegetables and fruits; beef and chicken; cooking oil and margarine; milk; egg; kerosene or natural gas for household-domestic use. The task to manage this state responsibility is entrusted to the State Logistic Agency (*Badan Urusan Logistik/ BULOG)*.[8] It is telling the absence from the list of staple food (rice, sago, corn), cassava (as raw food or processed: modified cassava flour).

Apparently, BULOG's task refers to guaranteeing the availability and even affordability of those nine basic commodities by controlling the supply and demand. The importance of this government intervention in the market of nine commodities should be understood in terms of the state interest in securing food sovereignty or security. Both concepts used interchangeable in existing laws-regulations and government policies. In any case, it is widely believed that disturbances in the market (availability and affordability) may lead to social-economic unrest, which threaten national political stability. The importance of securing the price stability of nine basic

---

[6] This system as to be implemented is found regulated in Art.38–43 Law 41/2009 (re. protection of Sustainable Agrricultural land (*Perlindungan Lahan Pertanian Pangan Berkelanjutan*)) and Government Regulation 12/2012 re. incentive to protect sustainable agricultural land *(Insentif Perlindungan Lahan Pertanian Pangan Berkelanjutan*).

[7] See: Ardiansyah Fadli, UU Cipta Kerja Izinkan Alih Fungsi Lahan Sawah, Ini Kriterianya, Kompas.com—22/02/2021. Cf. Sinuwun, Kebutuhan Lahan Untuk Pangan Capai 13,17 Juta Ha, https://pertanian.kulonprogokab.go.id/detil/10/kebutuhan-lahan-untuk-pangan-capai-1317-juta-ha. (14/12/2022).

[8] Established by Presidential Decree No. 114/U/KEP/1967 as amended several times, and lastly by Presidential Decree No. 3/2002 the Board has been disbanded. In replacement, the defunct board is resurrected as a general corporation (*perusahaan umum*) by Governemtn Regulation 7/2003 re. Establishment of the General Corporatiaon BULOG.

commodities is a hard lesson the Old Order (1965–1967) and New Order (1997/1999) government experienced. The insecurity of those nine commodities is one of the contributing factors leading to the government collapse.

A similar picture emerges on the unimportance of cassava in relation to the effort to seek alternative-renewable energy source (biofuel). Though outside the BULOG's scope of responsibility, the State (or other government agency) is entrusted with similar task: availability and affordability of energy source: natural oil and gas, coal and at present biofuel. The last named influences the support the government offers to big palm oil plantation companies throughout Indonesia. This fact may for a part explain the government dilemma of choosing—alternately—between maintaining subsidized natural oil and gas (for public and household use) and balancing the national budget by reducing or eliminating fuel subsidies. That said in the face of the undeniable fact that Indonesia national natural oil and gas reserve is declining fast, and Indonesia is now already a net importer of these energy source.[9] This fact may also explain the rush in recent years to explore the potential and development of another source of biofuel beside palm oil (a flexible crop), including Jathropha which unfortunately proved to be an utter failure.[10]

The successful processing and development of palm oil as a raw material for biofuel has encouraged the development of large-scale oil palm plantations at the outer islands: Sumatra, Kalimantan, Sulawesi to Papua. The express expectation is that sustainable production may continuously supply basic materials, which in turn may be processed as biofuel, either for domestic consumption or for export commodity, substituting or eliminating dependence to petroleum or coal. In comparison, cassava products or even its waste are not yet known, or popularized as was Jathropha in the past, to be a potential material for biofuel, produced on a large-industrial scale. The language used to describe cassava is as a crop having the potential or to be developed in the future for other multiple uses. For example, it is expressly stated that[11]:

> *Two types of biofuels being developed in Indonesia are: gasohol E-10, and biodiesel (B-10).* **Ethanol may be be procured by extracting cassava, a crop thriving all around Indonesia.** *While diesel oil is procured from palm oil, castor oil and coconut oil. Analysis conducted by BPPT stated that the market price of biodiesel B-10 is around Rp. 2.930 per liter, or Rp. 160, which is higher than the price of gasoline as subsidized by the government. The future advantage of producing biofuels is then the opportunity to reduce or even eliminate government subsidies, since addition of Rp. 160 may be acceptable.*

In support of the above argument, reference can be made to the ongoing research on the potency of cassava waste (non-consumables) as raw material for ethanol (biofuel). In one abstract of a scientific journal from 2019 is mentioned [9]:

[9] Press release of the Ministry of Energy and Minerals (*KEMENTERIAN ENERGI DAN SUMBER DAYA MINERAL, SIARAN PERS NOMOR: 028.Pers/04/SJI/2021*, dated 19 Januari 2021. As announced by the Minister: Indonesia natural oil reserve may last 9.5 years, and natural gas reserve may last a little bit longer: 19.9 years (from the present). To meet domestic energy consumption, Indonesia soon will become a net-exporter of non-renewable energy.

[10] Promode Kant and Shuirong Wu [8]. They wrote about and focus on India's experience. But the analysis and findings may well be applicable to Indonesia' experiment with Jathropa.

[11] Anonimus, Pengembangan Biofuel, April 17, 2018, https://www.bpdp.or.id/Pengembangan-Biofuel.

Cassave barks, as bio-waste, have the potential to be processed as energy source in the form of ethanol. Bioethanol may be procured through micro-organism fermentation. The result of which is glucose level of 9.9% with highest ethanol of 6.00 % if fermented 8 days.

The quotations above indicate the government low attention, at least at that period, 2018–2019, to cassava products as alternate staple food or raw material for biofuel. In other words, cassava, while being discussed in scientific circles as having the potential,[12] has not yet been taken seriously. In support of this observation, a quick peruse on legal materials or existing government projects or programs related to cassava, performed on line, showing minimal or casual attention to cassava as commercial crop, indicates the same. Is there any change to this attitude now?

5. Real and concrete threat of mass hunger or energy crisis?

Indonesia cannot but pay close attention to the impact of climate change, global warming, rising sea level, the potential extinction of various plants and endangered animals. Thus, not surprisingly the government decides to sign and ratify the Paris Convention.[13] The policies, rules, and regulation developed or coming out of the decision to become a party to this international treaty are mostly focused on reducing carbon emission and developing programs to reduce or contain deforestation. The focus on protecting the remaining tropical rainforest (and its endemic biodiversity) seems to be a rational choice given that these are considered common heritage of mankind. On the other hand, as a developing nation, Indonesia is adamant to utilize to the fullest its natural resources based on the nation's sovereign right to explore and exploit its natural resources.

The REDD and REDD+[14] are developed as G to G cooperation under the pretext of implementing parts of the Paris Convention. The basic idea seems to be that to reduce

[12] Montagnac et al. [10]. It is argued by the authors that: Cassava is a drought-tolerant, staple food crop grown in tropical and subtropical areas where many people are afflicted with undernutrition, making it a potentially valuable food source for developing countries. Cassava roots are a good source of energy, while the leaves provide protein, vitamins, and minerals. However, cassava roots and leaves are deficient in sulfur-containing amino acids (methionine and cysteine) and some nutrients are not optimally distributed within the plant. But compare with: Stephenson et al. [11].

[13] Law 16/2016, Ratification of the Paris Agreement to The United Nations Framework Convention on Climate Change (adopted 9 May 1992, enter into force: 4 November 2016). Today, 194 Parties (193 States plus the European Union) have joined the Paris Agreement. The Agreement sets long-term goals to guide all nations: substantially reduce global greenhouse gas emissions to limit the global temperature increase in this century to 2 degrees Celsius while pursuing efforts to limit the increase even further to 1.5 degrees; review countries' commitments every 5 years; provide financing to developing countries to mitigate climate change, strengthen resilience, and enhance abilities to adapt to climate impacts. See: The Paris Agreement, https://www.un.org/en/climatechange/paris-agreement.

[14] Reducing Emissions from Deforestation and Forest Degradation (REDD) dan REDD+ Reducing Emissions from Deforestation and Forest Degradation, role of conservation, sustainable management of forest and enhancement of forest carbon stocks in developing countries (REDD +). Periksa: Pertanyaan Seputar REDD+ dan Implementasi REDD+ di Indonesia, http://ditjenppi.menlhk.go.id/berita-ppi/33-beranda/1804-faq.html. Bdgkan: REDD+, http://ditjenppi.menlhk.go.id/kcpi/index.php/aksi/redd.

or stop carbon emission, states should cooperate and may trade quota. Arguably this kind of cooperation between developing and developed countries—in trading carbon emission quotas—are considered manifestation of the principle of common but differentiated responsibility.[15]

Paris Convention specifically addresses environmental problems stemming from global warming, climate change, rising sea levels, etc. But the impact or consequence thereof are not directly addressed. To know more about that, the list of global issues mentioned in the Millennium Development Goals and Sustainable Development Goals should be perused.[16] At the theoretical level, the solution to these global problems is said for nation-states to follow the precepts of sustainable development, a huge abstract and general idea found enshrined in the Stockholm Declaration 1972, Agenda 21, the Rio Declaration on Environment and Development, and the Statement of principles for the Sustainable Management of Forests.[17]

The option to adopt the basic principles of sustainable development, balancing the need to push economic growth with environment protections and other social-economic concerns, as a national policy or guiding principle, moreover, has for long been taken. The principles have been the guiding principle found in laws embodying national planning (The People's Consultative Assembly Decree on Broad Guidelines of State Policy, National-Provincial and District development plannings: short-, mid- and long term), basic laws relating to environmental protection, such as Law 24/1982 re. environmental management as amended by Law 32/2009 re. environmental protection and management), and the spatial planning law (Law 24/1992 as amended by Law 26/2007 and partially amended by Law 11/2020 re. Job Creation).[18] The last-named law, practically embodying the government policy on national economy, focusses not on the issue of hunger (food security or sovereignty) but more on the threat posed by unemployment-poverty due to Indonesian losing its competitive edge

---

[15] The concept of Common but Differentiated Responsibilities (CBDR) was enshrined as Principle 7 of the Rio Declaration at the first Rio Earth Summit in 1992. The declaration states: "In view of the different contributions to global environmental degradation, States have common but differentiated responsibilities. The developed countries acknowledge the responsibility that they bear in the international pursuit of sustainable development in view of the pressures their societies place on the global environment and of the technologies and financial resources they command." Similar language exists in the Framework Convention on Climate Change; parties should act to protect the climate system "on the basis of equality and in accordance with their common but differentiated responsibilities and respective capabilities." The principle holds that although all countries are responsible for the development of global society, each has a different set of capabilities that they can contribute to this project. See: POLICY BRIEF AND PROPOSALS: COMMON BUT DIFFERENTIATED RESPONSIBILITIES INTERNATIONAL MOVEMENT ATD FOURTH WORLD.

[16] See: United Nations, Transforming Our World: The 2030 Agenda for Sustainable Development, available at: https://sustainabledevelopment.un.org/content/documents/21252030%20Agenda%20for%20Sustainable%20Development%20web.pdf.

[17] Adopted by more than 178 Governments at the United Nations Conference on Environment and Development (UNCED) held in Rio de Janeiro, Brazil, 3 to 14 June 1992. Available at: https://sustainabledevelopment.un.org/outcomedocuments/agenda21.

[18] Law 11/2020 re. job creation law *(cipta kerja).*

in attracting foreign investment, that is compared with its close neighbors (ASEAN countries or within the Asia Pacific region).[19]

The above general law (addressing broad issues) may be compared to how the lawmakers perceived the issue of food availability. Law 18/2021[20] (rescinding Law 7/1996 re. food (*pangan*) have as its purpose securing food sovereignty, understood as the state's and nation's right to autonomously determine its food policy for the people and granting the people the right to determine their own food system, in accordance with local food potential. Thus, the end goal is food autonomy: the state and nation's ability to produce a variety of foods and attaining food security.[21] But then again, it should be noted that during the deliberation of the law before the parliament, the focus was more on: (a) meaning of food sovereignty, (b) position of imports to support food availability, (c) national food institution establishment, and (d) halal food issues.[22] The good news is that the above Law on food is going to be revised and the Parliament had already begun to recognize the need to amend said law in the context of capacity building of small- and medium-scale farmers.[23] Simply stated the law should be read and revised in conjunction with Law 19/2013 re.[24] protection and capacity building of farmers (*perlindungan dan pemberdayaan petani*), Law 13/2010[25] on Horticulture (*Horticultura*) and lastly Law 22/2019 re. agriculture system (*sistem budidaya pertanian*).[26]

This perspective on what to prioritize (economic growth above all else) may well be a dominant factor influencing how sustainable development (and Millennium Goals or SDGs) is understood and being realized in Indonesia. Regardless, the main purpose of this law apparently is to ease doing business by using a package of rules-regulation intended to deregulate and de-bureaucratize the government. But this law should be read in conjunction with 3 other laws (embodying government official policy on that field), i.e., empowerment of small-, middle-scale economic enterprises, taxation law and lastly the law, providing the legal basis for the development of the new capital city at Kalimantan. Arguably the last-named policy (relocate the capital city) is made purportedly to ease environmental pressure put so long on the current capital city (Jakarta) and Java island, which house approximately 80% of the Indonesian population on 20% of the whole land mass comprising Indonesia.

In contrast, the general threat of shortage of food, mass hunger, or other possible impact of global warming has not been responded with the making of nationwide policies

---

[19] Declared conditionally inconstitutional by the Constitutional Court by its judgment No. 91/PUU-XVIII/2020. The condition stipulated was that the Government within 2 years revised the Law following the formal procedure of law making. At the end of the two-year period, the President decides to circumvent this procedural barrier (going through the long process of bringing the draft through the parliament) by issuance of the Government Regulation in lieu of Law on Job Creation (*Peraturan Pemerintah Pengganti Undang-Undang 2/2002 tentang Cipta Kerja*).

[20] Law 18/2021 (rescinding Law 7/1996) re. food (*pangan*).

[21] Article 1 (2,3,4). Elaborated further in Government (implementing) Regulation 17/2015 on Food Security and Nutrition and GR 86/2019 on food security (keamanan pangan).

[22] Achmad Suryana et al. [12].

[23] Revised by the Parliament. See: DPR: LAPORAN PEMANTAUAN DAN PENINJAUAN UNDANG-UNDANG NOMOR 18 TAHUN 2012 TENTANG PANGAN: https://www.dpr.go.id/doksetjen/dokumen/persipar-Laporan-AKD-Buku-Laporan-Badan-Legislasi-DPR-RI-thd-Pemantauan-dan-Peninjauan-Undang-Undang-Nomor.-18-Tahun-2012-tentang-Pangan-1631075526.pdf.

[24] Law 19/2013 re. protection and capacity building of farmers (*perlindungan dan pemberdayaan petani*).

[25] Law 13/2010 on Horticulture (*Horticultura*).

[26] Law 22/2019 re. agriculture system (*sistem budidaya pertanian*).

or rules-regulations. The impression emerges that the one solution offered by the government is support continuing economic growth. That is not to say that the government, or at least academians, does not pay attention to the possibility of global warming resulting in threat to food security or sovereignty. For example, in the concern has been raised[27]:

The Dean of the Agriculture Faculty of GajahMada University-Yogyakarta, Ir Jaka Widada, MP, PhD, refers to FAO's statement reporting the possibility of mass hunger in 2050. This prediction is very real for the world and Indonesia. World population growth which at that year is expected to reach 10 billion people is one of the triggers. Another factor is climate change.

As to be read from the above quotation, climate change has not been considered serious enough a threat. What is more urgent, but it is predicted to be years away, is how to provide and secure food for the world-national growing population.

Arguably FAO does not fully focus on climate change but more on food security, which uses two different measurement: Prevalence of Undernourishment (PoU) and the Prevalence of moderate or severe food insecurity in the population, based on the Food Insecurity Experience Scale (FIES).[28] Both criteria are used as stick yard to measure the extent SDG2 goals has been meeting, i.e. a world without hunger, food insecurity, and malnutrition. Regardless, policies or programs developed in the context of attaining food security mention cassava more as a footnote. Note that FAO would be expected to spend too much time and effort in discussing the merits of cassava as the solution to world hunger.

This said considering that FAO official website,[29] expressly mentions the Country Programming Framework (CPF) which sets out four government priority areas to guide FAO partnership and support with the Government of Indonesia (GoI), i.e.:

Priority area 1. Support disaster risk reduction and improved resilience to climate change; Priority Area 2. Sustainable natural resource management for crops, livestock, forestry, and fisheries; Priority area 3. Improved agricultural productivity, value chain development and competitiveness and Priority area 4. Strengthen the enabling Policy, Legal and Institutional Environment to Improve Livelihoods, Food Security and Nutrition.

A different and more important issue is how both criteria as developed by FAO and other programs may be used and translated into food or agricultural policies at the national level. Reading the official website of the Ministry of Agriculture, a thought on a global strategy existing or to be followed in the future can be discerned[30]:

All agricultural development programs being carried out has as purpose not only to secure availability of food nationwide but is directed at improving farmers welfare [statement of Agung Hendriadi, head of the Food Security Agency (Badan Ketahanan Pangan]. The Minstry of Agriculture strategy for agricultural development and attaining national food

[27] Nograhany Widhi Koesmawardhani Pakar UGM Nilai 3 Negara Ini Siap Hadapi Bencana Kelaparan, Indonesia?—detikEdu, Rabu, 30 Nov 2022: https://www.detik.com/edu/edutainment/d-6435425/pakar-ugm-nilai-3-negara-ini-siap-hadapi-bencana-kelaparan-indonesia.

[28] Hunger and food insecurity, https://www.fao.org/hunger/en/.

[29] FAO in Indonesia, Programmes and Projects, https://www.fao.org/indonesia/programmes-and-projects/en/.

[30] Anonimus, Kebijakan Pangan Untuk Sejahterakan Petani, 12/10/2017 https://www.pertanian.go.id/home/?show=news&act=view&id=2290.

> *security is by implementing various breakthrough (programs), such as procurement of 180 thousand units of agricultural machinery and equipment, rehabilitation of existing irrigation networks covering 3.05 million ha, increase crop index, developing agricultural insurance (675 ha), increase meat production through SIWAB project (all female livestock should be made pregnant), construction of 3771 units of long storage/dam-irrigation trenches, procurement of superior seeds (rice, corn, soybeans, chilies, onions and others), controlling strategic food imports and stabilizing food prices through TTI (Toko Tani Indonesia; Indonesian Farmer Store). These policies (breakthroughs) has produced positive results and this serves as the basis for future policies intending to enable Indonesia, develop food estate and become the world bread basket in 2045 (lumbung pangan dunia).*

Aside from the difficulty of ascertaining whether policies mentioned above had been implemented with good result or are going to be implemented as breakthrough, what is striking is that all those goals (past or future) are not made based on calculation of PoU or FIES. The explicit strategy seems to be empowering small-scale farmers with the end goal Indonesia should, in the future, become the world bread basket. Except broad guidelines or programs found hidden between the wordings found in laws, rules and regulations issued by the government or various ministries, it is difficult to find any express reference to policies or programs made based on hard data as prescribed by the FAO. Or it may well be worded not in legal language.

It is telling though that the same FAO awarded the Indonesian government with: Acknowledgment for Achieving Agri-food System Resiliency and Rice Self-Sufficiency during 2019–2021 through the Application of Rice Innovation Technology. This proves again that in terms of food sufficiency, sustainable rice production is and continues to be prioritized.[31]

In any case, apparently, the issue of world or national hunger does not fall under the scope of authority of the Ministry of Agriculture. Instead, this issue and how to solve falls under the scope of responsibility of the Ministry of Health, as may be observed following news (including rules and regulations issued from time to time by the Ministry) on hunger and malnutrition. A specific public health issue is tackled by the Directorate General of Maternal and Child Health and Nutrition Development (*Dirjen Bina Gizi dan Kesehatan Ibu dan Anak/KIA*) Ministry of Health[32] or managed by cross sections—divisions of the local government as happened in Yogyakarta.[33] Mentioned that three different government agencies (health, agriculture-food security, and maritime and fishing) are to work together to manage the problem how to prevent hunger and malnutrition.

While it might not be fully correct, it can be safely assumed that at present there is no specific program, in the field of agriculture development or related to food security-sovereignty targeting cassava as prime agricultural commodity or proposing the development of this crop as alternate food source (substituting rice) or as product

[31] Anonimus, Berhasil Swasembada Beras, Indonesia Raih Penghargaan dari IRRIOleh Humas,14 Agustus 2022. https://setkab.go.id/berhasil-swasembada-beras-indonesia-raih-penghargaan-dari-irri/.

[32] See for instance: Herman/YUD, Kemkes: 80 Persen Penduduk Indonesia Kelaparan, 2015, https://www.beritasatu.com/news/298864/kemkes-80-persen-penduduk-indonesia-kelaparan. It is quoted: Directorate General of Maternal-Child Health and Nutrition Development (Dirjen Bina Gizi dan Kesehatan Ibu dan Anak, Minstry of Health (*Kementerian Kesehatan*), Anung Sugihanto argued that Indonesia at present face relative hunger. The disturbing fact is that only 20% of Indonesian population may secure the necessary 1.600 calory per day. Most of the Indonesian people only have access to or daily consume callories lower than this minimum.

[33] Sustainable Development Goals: Mengakhiri Kelaparan, http://bappeda.jogjaprov.go.id/dataku/sdgs/detail/2-mengakhiri-kelaparan.

industrially feasible to be processed into biofuel. These being important to determine whether the hope pinned on cassava (as flex crop) may be feasible realized, in contrast to what happened to the dream of cultivating Jathropha at industrial level to solve the problem of energy shortage.[34] It may also be possible that all the talk or discussion emerging about the potential of cassava, to end world hunger or alternate source of biofuel, will become a serial of government or private development projects—development projects to be sponsored by government loans and international donors. Or it may become a part of continuing government effort to do something good for society's welfare, the many failure of which described eloquently by Tania Li? That in consideration of the fact that cassava along with other edible plants (corn-palm oil, etc.) are being discussed, introduced, and promoted as flex crop. Meaning crop with multiple uses (food, feed, fuel, fiber, industrial material, etc.) that can be easily and flexibly inter-changed and therefore are attractive for investors because the flexibility allows investors to be flexible in deciding what to produce and sell based on price signals, vis à vis in diversifying markets for their investment while dealing with a single crop.[35]

6. Is cassava going to be upgraded: being potential to necessary and feasible?

At the same time, a significant trend emerges indicating the rising attention given to the future potential of cassava (especially its tubers) both as a food source (staple food) and more importantly as raw material to be processed into biofuels or for other industrial needs (processed into ethanol).[36] In response to the positive reports on the potential of sorghum and especially cassava as food source (lessening dependency on rice), the Minister of Defense, Prabowo Subianto, proposed, at the Global Food Security Forum in Nusa Dua Bali, Sunday (13/11), the development of cassava as main food commodity of the future. The question here is then why should the Indonesian Defense Minister proposed this idea, what cassava to do with defense issues? A quick answer to this kind of queries was offered by a political and intelligence communication observer, Susaningtyas Nefo Handayan. She argued that[37]:

> *Food security is a key component of the People's Defense and Security System (system pertahanan dan keamanan rakyat semesta). The history of the struggle for the independence of the Republic of Indonesia proves that, in each battle to do so, logistical resilience is a determining factor.*

She further argues that therefore it was a right decision of the President to appoint the Ministry of Defense as the leading sector handling the issue of food crisis. The incumbent Minister of Defense in that context stated[38]:

[34] See: Hengky Wijaya [13].

[35] Hengky Wijaya ([13], p. 1).

[36] Anonimus, Sorgum & Mocaf: Perkuat Ketahanan Pangan, Trubus, 636, November2022/LIII.: 9–13. Cassava harvest period is 6–12 months and around 500,000–1,000,000 ha of land is needed for cassava and sorghum to be able to function as alternative substitutes for wheat (averaging 10 million tons per year.

[37] Anonimus: Dukung Pernyataan Menhan Prabowo, Pengamat: Ketahanan Pangan Bagian dari Sishankamrata, 17 november 2022, https://indonesiadefense.com/dukung-pernyataan-menhan-prabowo-pengamat-ketahanan-pangan-bagian-dari-sishankamrata/.

[38] Anonimus, Bahas Singkong di G20, Prabowo: Indonesia Bisa Sediakan Pangan untuk 8 Miliar Orang di Dunia, 15 november 2022, https://headtopics.com/id/article-headline-31697658.

> *Our goal is (for Indonesia in the future) to be able to feed 8 billion people (...) the government's task is to ensure the availability and affordability of food to achieve the target of zero hunger. This target is mentioned as number two in the Sustainable Development Goals (SDGs).*

The Minister of Defense, Prabowo Subianto, before the G20 forum held in Bali (2022) also confidently asserted that[39]:

> *(...) cassava is the most efficient crop as it could produce 250,000 calories while only needing 65 cubic meters (cbm) of water per tonne, which was far less than rice, which needed 1139 cbm, wheat needing 954 cbm and maize needing 815 cbm. (...) Indonesia can become the foremost producer of cassava. [...] Cassava is now a strategic food crop. (...) Indonesia is currently producing instant noodles and pasta from cassava. On top of that, cassava could also be processed into bioethanol, alcohol, vitamins, bioplastics, glue, explosives and cattle feed while being 100 percent gluten-free with a low glycemic index, high in iron and calcium.*

The above idea arguably should be read in conjunction with the incumbent President's plan to make Central Kalimantan (Kapuas and Pulang Pisau) the bread basket of Indonesia intended to replace Java which lost this position as this island experience massive loss of agricultural land due to rapid urbanization and industrialization. Available land in Kalimantan, considered to be still in abundance and underdeveloped, should be cultivated as food estate, strengthening the national food security in the context of national sovereignty. The development plan or program is designed as a cooperation between the Ministry of Defense, Ministry of Public Works and Housing (PUPR),[40] Ministry of Agriculture, Ministry of Environment and Forestry and Ministry of State (owned) Corporations (BUMN). The Ministry of Environment and Forestry has the responsibility to assure the smooth harmonization of this program with other programs purporting to restore wetlands.[41] As for the reason why the President appoints the Minister of Defense as the program coordinator[42]:

[39] As reported by Deni Ghifari, Tempeh, cassava "the answer" to staple-food insecurity, the Jakarta post, Nov. 15, 2022, https://asianews.network/tempeh-cassava-the-answer-to-staple-food-insecurity/.

[40] See: Government Regulation (*peraturan pemerintah*) 20/2006 re. irrigation. In this regulation, we can find the statement that food security should be realized by maintaining a sustainable irrigation system, from development to operation, service irrigation networks. This task is entrusted to the Public Work and Housing Ministry (*Pekerjaan Umum dan Perumsahan Rakyat/PUPR*).

[41] Anonimus, Menhan didampingi Wamenhan Temui Menteri LHK Bahas Kelanjutan Program Food Estate, WIRA melalui Edisi bulan Juli–Agutus 2020 Nomor 2, available at https://www.kemhan.go.id/wp-content/uploads/2021/06/WIRAbaru2020INDkecil.pdf. As for wetlands restoration, the note to be made is that efforts to restore Indonesia's tropical peatlands have been accelerated by the establishment of the Peatland Restoration Agency in early 2016 by Government Regulation 1/2016. The restoration action policy includes the rewetting, ion, and revitalization of local livelihood (known as the 3Rs). At the ministerial level, the restoration of peat ecosystems is regulated by the Ministry of Environment and Forestry Regulation Number 16/2017, which set out the technical guidelines for restoration of peatland ecosystem. It is intended to provide technical guidance for the national government, regional/provincial governments, communities (including indigenous people), and those responsible for businesses and or activities in restoring the function of peat-mangrove ecosystems. Tri Wira Yuwati et al. [14]. For further information on the Agency, visit: https://brgm.go.id/.

[42] As quoted from Chandra Gian Asmara [15].

> *Defence does not only cover the responsibility of procurement and managing the availability and readiness of weaponry for the armed forces (alat utama system senjata tentara nasional Indonesia, alutista), but also food security.*

Given that response, arguably what has been the basis for this policy (developing food estates) is not particularly the impact of harvest failure, mass hunger, and malnutrition from lack of food, all possible consequences of climate change, but something other issues, considered significantly more important. Possible cause or background for the above policy is the government plan to move the capital city from Jakarta to an area straddling two districts: Kutai Kartanegara and Penajam Paser Utara (both are situated at East Kalimantan).[43] To guarantee the success of this huge project, around the new capital city, expected to ease the environmental pressure borne by Java, the government at the same time must prepare food estates, to provide for the future residents populating the new capital. However, food estates are also being prepared and developed at other islands, with success or sometimes ending in failure.[44] Regardless, the basic assumption driving this program (and other government programs related to conversion of forest land) seems to be the general perception that forest (despite recognized importance in terms of environment, climate change prevention, etc.) are considered undeveloped and underutilized. Forest land should and needs to be converted to other more beneficial use for mankind or society's welfare. The result is increasing rate of deforestation as the table for the period of 2017–2021 shows (**Table 2**)[45]:

In line with the numbers above, the Central Statistic Bureau (Badan Pusat Statistik (BPS)[46] reported a loss of forested land amounting to 956.258 hektare (ha) (0.5% of Indonesian land mass) during 2017–2021—deforestation occurred in Kalimantan, Papua, and Sumatra. Increase of forested area is reported in Bali-Nusa Tenggara, Sulawesi, Jawa, dan Maluku, but the increase is slower than total loss of forested areas. The main cause of this deforestation is the government policy of supporting large-scale investment in palm oil plantation or conversion to other more lucrative enterprises, such as mining of precious minerals (coal, gold, nickel, etc.). Arguably, continuing deforestation is also justified by the so-called need to establish food estates, producing mostly rice.

In any case, what is even more striking is that even in the above grand scheme of relocating the capital city and in support of that effort developing new food estates (and the justification of deforestation of the area), cassava was never explicitly mentioned—only in passing and that only about its potential or possible development. A similar impression of the relative unimportance of cassava also emerges

[43] Law 3/2021. Government official website containing all necessary information or at least for public information regarding the development of the new capital city (*Ibu Kota Negara*) is: https://ikn.go.id/en/letak-ibu-kota-baru-indonesia-bernama-nusantara-ini-detail-lokasinya. It is also mentioned that Law 3/2021 will soon be revised. See: Rofiq Hidayat [16].

[44] Abdul Basith Bardan [17]. Cf. V. Arnila Wulandani, [18].

[45] Rate of Deforestation; Anonimus, Pengurangan/Penambahan Luas Tutupan Hutan Indonesia (2017–2021): https://databoks.katadata.co.id/datapublish/2022/12/21/luas-hutan-indonesia-berkurang-hampir-sejuta-hektare-dalam-5-tahun. Cf. anonimus, "Data KLHK Tahun 2022 Periode I: Hutan Primer Berkurang", https://www.cnnindonesia.com/nasional/20220413073537-20-784096/data-klhk-tahun-2022-periode-i-hutan-primer-berkurang.

[46] Adi Ahdiat, Ini Luas Tutupan Hutan Indonesia, dari Sumatra sampai Papua, https://databoks.katadata.co.id/datapublish/2022/12/20/ini-luas-tutupan-hutan-indonesia-dari-sumatra-sampai-papua.

No	Name (of the main island in the archipelago)	Amount/hectares
1.	Kalimantan	−654.664
2.	Papua	−610.405
3.	Sumatera	−310.374
4.	Maluku	78.088
5.	Java	113.884
6.	Sulawesi	202.057
7.	Bali-Nusa Tenggara	225.156

Table 2.
Deforestation rate.

when tracing the wordings of the basic laws on food (Law 18/2012), food & nutrition security (Government Regulation 17/2015) or other laws and regulations pertaining to spatial planning (Law 26/2007), environment management (Law 32/2009), etc., which mention or become the basis for policies in the agricultural field or government programs related to establishment of food estates. The same could said when perusing the package of top down development programs embodied in laws and regulations, and other sectoral (forestry, agriculture, industry etc.) development programs.

Instead what is to be found is rather negligible, a passing consideration of cassava future potential. It is the Ministry of Agriculture who promoted cassava as local strategic food commodity. In this context, the Minister mentions the existence of a pilot project to establish—develop an integrated cassava estate.[47] In a similar vein (Zulkifli, Director of PT Permata Agro Utama) argues[48]:

> *Cassava should not be underestimated. Within the next 25 year, it may become a strategic product. Cassava possess many advantages, it may reduce Indonesia's dependence to wheat import, and may be used as raw material for pharmaceutical products, whereas as of today, Indonesia imported 96% of those materials. The society expect that the government develop cooperation programs, offering guidance to society and develop unproductive land into productive.*

In the same news, the Director General Food Crops, Ministry of Agriculture, Suwandi, is reported to state that there are already circulated plans to develop cassava as part of the food diversification program at the local level. In realization of this plan, the Ministry had it performed a mapping project of cassava centers [small-scale plantations]. In 2019, it is recorded a total area of 628 thousand hectares, producing 16.35 million ton. Unfortunately, in 2022, there are no traces left of this ambitious plan or how it might have been developed.

Briefly stated, after perusing the existing laws and regulations, or looking at existing policy statements on development in general or cassava in particular—in terms of food security/sovereignty or in other context—arguably up to present there

[47] Anonimus, Kementan Bersama MSI Menjadikan Singkong Pangan Lokal Strategis, tanpa tanggal-tahun https://www.pertanian.go.id/home/?show=news&act=view&id=4412. & Olahan Singkong Mampu Tingkatkan Nilai Jual Bernilai Ekonomi Tinggi, https://www.pertanian.go.id/home/?show=news&act=view&id=4869.

[48] Ibid.

is no indication that the national government is really putting effort in the realization of so far fledging proposals related to cassava improvement or development as flex crop. This is said in consideration of some local government initiatives, as in providing (financial or other kind of) assistance to small-scale cassava farmers at Bangka-Belitung.[49] But this more concrete and real effort and attention to local farmers planting cassava apparently is limited to local (district) governments only. Such initiatives have yet to be developed further, only the condition if feasible, into national program and related to the effort of ending world hunger, battling climate change or lessening dependency on oil and gas as energy source.

7. Conclusion

Apparently, rice is still considered Indonesia staple food to which sustainable production must be guaranteed. The prevailing law and regulations, symbolic expression of development priority, still reflect this bias. In contrast, at the national level, cassava's advantages and potential are still only discussed at scientific circles only. The promise of cassava as flex food, lessening Indonesia dependence on rice and possible development of cassava waste as biofuel has yet to be developed into feasible policies or development programs. The realization of which would then endanger various development programs, funded by the government or to be sold as a commodity of development idea with the best of intentions.

Development and enhancement of cassava, at present being produced, often incidentally to be consumed locally, into a commodity, produced and processed at the industrial level, has yet to materialize. Discussion of the potential of cassava, either as food crop or as raw materials for other uses, has yet to reach the level of being hype, as being experiences by Jathropha a decade earlier. While it is true that there is trace of hope or in Tania Li's language the desire (wish) to make cassava a super commodity to solve multiple problems: from hunger, food security to energy scarcity, the efforts to do so is at most at a preliminary stage.

[49] Anonimus, Tingkatkan Pendapatan Petani Singkong Babel, Kementan Beri Bantuan Saprodi. https://www.pertanian.go.id/home/?show=news&act=view&id=4570.

Author details

Tristam Pascal Moeliono* and Koerniatmanto Soetoprawiro
Faculty of Law, Parahyangan Catholic University, Bandung, Indonesia

*Address all correspondence to: tristam@unpar.ac.id

References

[1] Koerniatmanto S. Pengantar Hukum Pertanian. Bandung: Gapperindo; 2013

[2] García-Villegas M. The symbolic uses of law: At the heart of a political sociology of law. In: In the Powers of Law: A Comparative Analysis of Sociopolitical Legal Studies (Cambridge Studies in Law and Society). Cambridge: Cambridge University Press; 2018. pp. 19-37. DOI: 10.1017/9781108584975.003

[3] van Klink B. An effective historical view on the symbolic working of law. In: Zeegers N, Wittteven WJ, van Klink B, editors. Social and Symbolic Effects of Legislation under the Rule of Law. Lewiston, NY: Studies in Political Science; 2005. pp. 113-144

[4] Murray LT. The Will to Improve: Governmentality, Development and the Practice of Politics. Durham-London: Duke University Press; 2007

[5] Vel JAC. How could the jatropha sector live on subsidies and green investments without producing oil or biodiesel? In: Vel JAC, Simandjuntak D, editors. JARAK, the Short History of Jatropha Projects in Indonesia. IIAS. book chapter: Leiden; 2014

[6] Vel JAC. Trading in discursive commodities: Biofuel brokers' roles in perpetuating the Jatropha hype in Indonesia. Sustainability. 2014;**6**(5):2802-2821. Article in journal: refereed

[7] Vel JAC. What are the benefits of analyzing jatropha activities as a commoditization process? In: Vel JAC, Simandjuntak D, editors. JARAK, the Short History of Jatropha Projects in Indonesia. IIAS. book chapter: Leiden; 2014

[8] Kant P, Shuirong W. The extraordinary collapse of Jatropha as a global biofuel. Environmental Science & Technology. 2011;**45**(17):7114-7115. DOI: 10.1021/es201943v

[9] Widyastuti P. Pengolahan Limbah Kulit Singkong sebagai Bahan Bakar Bioetanol melalui Proses Fermentasi. Jurnal Kompetensi Teknik. 2019;**11**(1):41-46

[10] Montagnac JA, Davis CR, Tanumihardjo SA. Nutritional value of cassava for use as a staple food and recent advances for improvement. Comprehensive Reviews in Food Science and Food Safety. 2009;**8**(3):181-194. DOI: 10.1111/j.1541-4337.2009.00077.x

[11] Stephenson K, Amthor R, Mallowa S, et al. Consuming cassava as a staple food places children 2-5 years old at risk for inadequate protein intake, an observational study in Kenya and Nigeria. Nutrition Journal. 2010;**9**:9. DOI: 10.1186/1475-2891-9-9

[12] Suryana A, Khalil M. Proses dan Dinamika Penyusunan Undang-Undang Nomor 18 Tahun 2012 tentang Pangan. Forum Penelitian Agro Ekonomi. 2017;**35**(1):1-17. DOI: 10.21082/fae.v35n1.2017.1-17

[13] Wijaya H. Deconstructing a Biofuel Hype the Stories of Jatropha Projects in South Sulawesi. Indonesia: Dissertation, Leiden University; 2018

[14] Yuwati TW et al. Restoration of degraded tropical peatland in Indonesia: A review. Land. 2021;**10**(11):1170. DOI: 10.3390/land10111170

[15] Asmara CG. CNBC Indonesia, Bukan Mentan, Ini Alasan Jokowi Pilih Prabowo di Food Estate. 2020. Available from: https://www.cnbcindonesia.

com/news/20200713184745-4-172363/bukan-mentan-ini-alasan-jokowi-pilih-prabowo-di-food-estate

[16] Hidayat Rofiq, Revisi UU IKN, Ada Persoalan Serius dalam Tata Kelola Legislasi. 2022. Available from: https://www.hukumonline.com/berita/a/revisi-uu-ikn--ada-persoalan-serius-dalam-tata-kelola-legislasi-lt63804704b727d/

[17] Bardan Abdul Basith, Pengamat jabarkan deretan kegagalan food estate yang dicanangkan pemerintah, Senin. 2020. Available from: https://nasional.kontan.co.id/news/pengamat-jabarkan-deretan-kegagalan-food-estate-yang-dicanangkan-pemerintah

[18] Wulandani VA. Food Estate dan Nasib Keanekaragaman Hayati di Tanah Papua. 2021. Available from: https://econusa.id/id/ecoblog/food-estate-dan-nasib-keanekaragaman-hayati-di-tanah-papua/

Chapter 2

Assessing the Traditional Food Processing Economy of Agrarian Communities in Southwestern Nigeria: The Case of Esan Women and the Cassava Processing Industry

Emmanuel Osewe Akubor, Beatrice Amili Akubor and Funmilayo Evaristus Awoyera

Abstract

Extant studies have established that in Africa (with specific reference to Esan people, southwest Nigerian area), the societies from time immemorial developed indigenous or traditional methods/ways of processing farm products to ready-made food for consumption. Analysis shows that one of such was the processing of cassava (*Manihot esculenta*) to various food items, which include *Gari/garri* (Cassava flakes), *Akpu/fufu* (fermented cassava product), *Bobozi/Tapioka* (Cassava chips), *Elubo/Lafun* (Cassava flour) amongst others. In all this process, the women folks are the ones that are largely involved. This chapter is thus an attempt at interrogating the processes involved in the production of one of the most popular sources of carbohydrate *Gari* from Cassava amongst Esan agrarian societies of southwest Nigeria. This chapter tries as much as possible to interrogate how the women using local technology and fabricated devices are able to process and produce cassava related food for both family consumption and commercial purpose. The qualitative method is used for this particular research. Data obtained from primary and secondary sources were deployed to carry out the study with an analytical and narrative historical approach. This includes historical, descriptive and analytical approaches based on gathered evidence. The result shows well outline methods adopted in producing this food item so as to make sure it is nutritious, marketable and can be preserved using local technology.

Keywords: food production, processing, Gari, cassava, Nigeria

1. Introduction

Experts have argued that as far as the history of human existence is concerned, food production/processing is the most widespread form of human activity.

In most of the Developing countries of the world (Nigeria inclusive), indigenous (also referred to as traditional) food processing had started from early times (pre-colonial period) and most of the traditionally processed foods are based on a combination of preservation technologies, common sense and indigenous technology [1]. Akubor and Akubor [2] specifically noted that it is the main-prolong of the economic activity predominantly subsistence farming, providing a livelihood to more than three-quarters of the human race. The availability of much process from this agricultural practice, initially led to a situation in which in the rural areas, the traditional pattern of food utilisation involves mainly the consumption of different food items in the fresh state with little or no post-harvest processing, before being cooked into meal. However, with the advancement of most societies, as well as the increase in population, there arose the need to advance method that will help preserve agricultural produce for the rainy days. It was this situation that led to development of system of Food Processing.

Generally, the food industry can be broadly classified as consisting of the large-scale foreign-based multinationals with local subsidiaries, public sector-backed production outfits (parastatals or state-promoted companies), medium-scale indigenous and foreign operators, small-scale indigenous operators and the cottage or sole proprietors outfit. Whilst the chapter recognises that all these sectors are crucial for the economic development of any society, however the paper is more concerned with small-scale indigenous (Indigenous) Food Processing operators which are dominant in the case of Nigeria.

As a way of clarification, the small-scale indigenous (traditional) food processing operators are partly urban-based, whilst the majority is based in the rural areas as cottage level producers. Available evidence indicates that this group engages in the processing of various local stables that are derived from roots and tubers, cereals, fruits and vegetables, products of animal origin such as dried and smoked fish, meat, local cheese and food additives such as flavouring condiments and other food seasoning materials.

One unique feature of this group is that despite being unskilled, within years of practice they become specialised in their trade and have over the years established inter and intra-regional professional groups, maintaining contacts with and between various ecological regions. This is the case of the Esan Women in the Cassava (Food) processing industry. Although a completely agrarian society (dotted with few industries), southwest Nigeria, the women have wholly dominated the industry, giving rise to, and increased, economic specialisation. It is in line with this that there is the urgent need to locate the area and people in the global map.

2. The Esan area and people: a geographical conceptualisation

By Definition, the Esan people, and the land (Oto Esan as it is described by the people) is an Edo speaking territory, which belongs to the Kwa branch of the Niger–Congo family of languages. Located in southwest Nigeria (West Africa), the territory and people (which is largely agrarian) are bounded to the Northwest and North by the Ivbiosakun and Etsako section of Northern Edo, to the West and South-west by the Benin Kingdom, to the South and South-east by Western Igbo and to the East by the River Niger and the Igala people. In a more specific term, Ukaugo and Akubor [3] argued that Esanland is located in the present Edo State of Nigeria and Esan language is spoken by the people. The culture and origin of the people is linked with the Benin. It is largely in the thick forest region with boundaries with the Etsako, Owan, Benin,

Aniocha and the Ika. Esan was corrupted to Isan, Esa and Ishan by the Europeans and has been used interchangeably ever since to represent the people, their language and culture. In Esan land, Agriculture is the main occupation of the people, with the men majorly cultivating cassava, whilst the women cultivate cassava and little vegetable farms. Esan women do some petty trading and also weave clothes [4–6] had identified the major markets around the area as those trading stations on the Niger, the most considerable of which is Illushi, an evacuation point for much produce from Esan land (including Garri and other cassava related processed food) to neighbouring states [4].

Presently, it is estimated that the Esan people who reside in Esanland number about from 1.5 million to 3 million citizens in Nigeria and there is a strong Esan diaspora ([3], p. 15).

3. The indigenous (traditional) food process industry in Nigeria: a brief background

Traditionally, the food processing industry in most African states is controlled by the women folks. This is because whilst the production of the food is seen as a man's job, because of the hectic nature of farming, it is argued that the women compliment the efforts of the men by converting this food items from their raw stage to processed items for both internal consumption and trade [7, 8]. According to Henn [9], women around the area are known to have long supplied the bulk of labour required in food production and processing. Related researches have also established that women in our area of studies, are as hard working as the men folks when it comes to agriculture and food processing, since they are fully involved in clearing of land, root crop production, planting, weeding, harvesting, transportation and processing, store and marketing of the final product ([10] cited in [2, 11]).

Under the traditional Esan system, although women were not allowed to own large farmlands, they however, owned and cultivated plots belonging to their husbands, where they planted melon and related vegetable crops, which served as supplementary to the family's food table [12]. In additions, they were allotted small plots close to their husband's farm where they cultivate cassava and cocoyam. Whilst the cocoyam is directly boiled/pounded and eaten; the cassava is processed either for *garri, Tapioka* or *akpu*. At the beginning of the farming season, apart from providing food and water, for those working in the farm, women also assisted in sowing, planting and harvesting, and in marketing of the harvested farm produce, at the end of the farming season [13]. Throughout the entire cassava production and processing, Esan women typically carry out 70 percent of the work, including; planting, weeding, harvesting, transporting cassava, peeling, grating and/or soaking, bagging and selling.

4. Cassava processing industry: production, processing and source of labour

Amongst the people of Esan, cassava is processed into several stable food products. One of them is the processing cassava into granules known as garri. *Garri* is a menu to average Nigerian and before the current hardship and inflation in Nigeria was considered a food for the poor. Garri can be eaten raw (in its dry powdered state), or soaked in water with groundnuts, beans, palm kernel, coconut, bean-cake. Garri could be taken alone or added into beans porridge or made with boiled water into a solid meal of *eba* and taken with all kind of vegetable soup. Apart from yam (which is the major tuber crop cultivated mainly by men), cassava tuber is considered the most important tuber crop amongst the farming communities of southern Nigeria in generally and Esan communities in particular ([14]:190–192, [15]:1) Cassava could also be dried and processed into powder for making paste which is eaten with soup. In some other instance, it can be boiled, soaked for some time and sliced into straight wet chips and consumed with either coconut or groundnut (Bobozi/tapioca/abacha) [16].

Commenting, Oyewole [17] argued that the processing of cassava for food involves combinations of fermentation, drying and cooking. In the view of the source, Fermentation is an important method common in most processing. According to Oyewole [17], there are many fermentation techniques for cassava, which can be broadly categorised into solid-state fermentation and submerged fermentation. Commenting further, the source opined thus:

> *Solid-state fermentation, typified by gari production, uses grated or sliced cassava pieces that are allowed to ferment while exposed to the natural atmosphere or pressed in a bag. Submerged fermentation involves the soaking of whole peeled, cut and peeled, or unpeeled cassava roots in water for various periods, as typified by the production of fufu and lafun in Nigeria. Traditionally, cassava is fermented for 4 to 6 days in order to effect sufficient detoxification of the roots.*

In line with the above and considering the hectic nature involved in the processing, it is therefore not surprising that the traditional (indigenous) Esan cassava processing industry witnesses a marked division of labour, with the women doing a large chunk of the work. In most cases, the adult male assisted the women in the harvesting of the cassava from the farm. From this point, the women assisted by their children take over the preparation process of the cassava starting with the collecting of the harvested cassava tubers from the farmland to the house where process begins ([12, 18]:23–50, [13, 19]). The job of preparing/processing cassava was really a tedious one, requiring an enormous labour input. The labour requirement was met by the women and children in the household. In cases where the quantities of cassava tubers to be prepared are much, extra labour might be desired, especially when it is for commercial purpose as distinct from production for family consumption. In this situation, the women would mobilise their friends or a communal work group involving a desirable number of relatives/age-group peers to help at every stage of Cassava processing. The contractual agreement stipulates how the labour expended would be repaid.

5. Cassava: the plant, planting and harvesting

Cassava is an important food crop in the tropics and many countries in Africa. Cassava has been found to be a major crop in arable cropping of smallholder farmers [16]. The crop contributes significantly to the diets of over 800 million people, with per capita consumption averaging 102 kilogrammes per year. In some areas of Africa it constitutes over 50 percent of the daily diets of the people. Amongst the farming communities in Nigeria, cassava production is well-developed as an organised agricultural crop. It has well-established multiplication and processing techniques for food products and cattle feed. There are more than 40 cassava varieties in use. Planting occurs during four planting seasons in the various geo-ecological zones. Cassava is grown throughout the year, making it preferable to the seasonal crops of yam, beans or peas.

It displays an exceptional ability to adapt to climate change, with a tolerance to low soil fertility, resistance to drought conditions, pests and diseases, and suitability to store its roots for long periods underground even after they mature. Use of fertilisers is limited, and it is also grown on fallow lands. Harvesting of the roots after planting varies from 6 months to 3 years. The land holding for farming in Nigeria is between 0.5 and 2.5 hectares (1.2–6.2 acres), with about 90% of producers being small-scale farms [16].

Historically, Cassava and its related products constitute a major staple food amongst the populace in Nigeria. Cassava (a perennial woody shrub with an edible root) was first introduced to Nigeria in the sixteenth century. With its introduction, it gained popularity amongst the rural farming societies of southern Nigeria, where the agro-ecological zone favourably supported the cultivation of such crop. However, over the years, the majority of cassava farmers cultivate small farm area which is not conducive or economical for mechanisation. Despite these challenges, cassava is one of the fastest expanding staple food crops in cassava consuming countries and has continued to gain prominence amongst farmers whilst the industrial demand is also rising consistently [20]. FAOSTAT [21], has established that as it 2019, Nigeria produced about 60 million tonnes. Despite being the largest producer of cassava in the world, more than 90% of cassava produced in Nigeria are consumed locally [22].

6. Stages of cassava processing

Traditionally, cassava is processed before consumption. Processing is necessary for several reasons. First, it serves as a means of removing or reducing the potentially toxic cyanogenic glucosides present in fresh cassava. Second, it serves as a means of preservation. Third, processing yields products that have different characteristics, which create variety in cassava diets. The processing of cassava began immediately after the harvest of the tubers with its peeling, washing, grating, pressing (dewatering), sifting, frying, colouring, drying, packaging and preparation for eating in various forms.

7. Peeling and grating stage

Cassava processing commenced with it peeling. Peeling involves the removal of the outer layer with a knife by women and children. The peeled cassava is washed with clean water. This follows the grating of cassava tuber into paste form to make it ready for pressing i.e. extraction of its liquid content. Before the fabrication of modern machine, the grating was done by scrubbing the cassava tuber against flat metal sheet on which is made sharp holes. An optional stage in cassava was the fermentation of the grated cassava paste. After grating the tubers, it is usually packed in basket, locally woven bags, or bowels. It will be allowed for 2 or 3 days before it is subjected to pressing to achieve dewatering. When cassava is subjected to this process it produces a sour taste different from those not subjected to the process, but this is essentially based on choice.

8. The stage of dewatering (pressing to remove water)

The pressing stage followed after grating, the grated cassava (which is now in its semi-liquid state) is packed in bag(s) and subjected to a pressing device. The bag of the grated cassava is tied on a device, whose grip was tightened at regular intervals until it reaches reduction in the liquid content. In those days, the pressing device was made from bamboo and locally woven ropes but nowadays it is done with the aid of a machine. In some other area, the bag of grated cassava is placed in between two large stones or heavy objects to achieve the same aim of extracting the liquid as much as possible from the cassava. Whilst it was obvious that dewatering was done to achieve solidifying of the cassava paste, the reason for dehydration might have been related to how it eases the process of frying. It has been argued that the process of fermentation while dewatering helps to detoxify cyanide in cassava. It is important to note that the water extracted through this process is not completely useless as it can also be preserved for making starch for clothes or consumption as common amongst the Urhobo and Ijaw people of the Niger Delta area of Nigeria [23].

9. Sifting stage of processing

After the pressing/liquid extraction stage, sifting followed immediately. This stage involves the separation of the finely grated part of the cassava paste suitable for consumption as the end product from the chaffs and tiny lumps that might have found their way into the whole of the grated paste through the use of the locally made sieve. This is done by packaging a measured quantity of the grated cassava, which assumes the form of flour after dehydration into the sieve. Sifting is achieved by gently running the palms through the measured quantity of cassava flour on the sieves. The portion that went through the bowl placed under/in between the laps is considered fit for further processing into the end product. Colouring constituted another optional activity of the stages involved in the processing of cassava. It is the colouring activity that gave rise to the two colours of *garri* (white and red) that is available amongst the people. White is the natural colour of *garri*, whilst red (although described as red, it is actually yellowish in colour) is the colour produced when red oil is added to it. Amongst the professionals, the colouring is believed as a matter of opinion that the application of palm oil as a colouring agent improves the quality of the *garri* and reduces whatever negative effect that might result from its consumption. According to Eboyehi [24], the colouration also helps to reduce the acidic nature, that could be harmful to some consumers. Thus the red *garri* is generally preferred by the people, particularly in the preparation of *eba*, a solid meal prepared from *garri* is commonly eaten amongst the southern part of Nigeria.

10. Frying and sun drying

The last stage in the processing of cassava into *garri* is the frying of the sieved cassava flour. This involves the heating of the flour in an earthen or aluminium pot over a fire made to suit the purpose. Whilst the heating is on, the processor is expected to continuously undertake the turning of the cassava flour until it is certified dried. *Garri* may be heated enough for consumption on fire, whilst the process of drying was completed by further spreading out of the *garri* in the sun. Sun drying the *garri* was optional and in most Esan processing communities discouraged and frowned at. In fact the sun drying method was always seen of shortchanging people and people are discouraged from consuming such product except for commercial purpose. Frying is however considered better, because it is neater and aid preservation. The sun drying method is not popular amongst the people probably because of the effect on the nutritional content of the food. According to observation, it has been opined that sun drying causes large losses in carotene content, Vitamin A and Vitamin D. Nutritionists are of the view that retention of vitamins in dehydrated foods is superior to sun dried foods. They have also opined that rancidity is an important problem in dried foods. According to this school of thought, Enzymatic browning or caramelisation types of reaction may occur and can be controlled by addition of Sulphur dioxide [23].

It is important to note that apart from the production of garri from cassava, the Esan women have been able to process other food items from same source. For example there is the Cassava Foo-Foo (*akpu*), which is derived through whole or cutting cassava roots are peeled washed and allowed to ferment naturally to a soft texture in 4–5 days. This is cooked into foo-foo (paste) and eaten with vegetable soup. There is also the tuber flour (*lafun/Elubo*), which is gotten through clean cassava root cut into chips, allowed to ferment 1–2 days, sundried, blended and sieved. The flour is cooked into a paste with water and eaten with vegetable soup. There is also the *Bobozi/Tapioka*, which is cassava tuber, peeled and boiled, after which it is sliced into chips and soaked in clean water (overnight to reduce the level of acidity). This is eaten with coconut, palm Kernel, groundnut or specially prepared sauce.

11. Conclusion

From the discourse, it is clear that the food processing industry of the people (as most African societies) has traditionally championed by the women using local technology. The product of this process has over the years geared towards both family consumption and exchange. This product has also attracted high purchase in both local, inter and intra-regional markets around the area. In fact in the Illushi market which is the biggest market in Edo state (linking three other states- Kogi, Anambra and Delta states of Nigeria), the processing and trading in Garri has been considered so were essential and important, resulting in the establishment of a separate section for the trade within the market. Over the years, this has attracted contact between various ecological regions, differential needs and trade [13, 25, 26]. This gave rise to, and increased, economic specialisation especially in the processing industry. Scholars have argued that this specialisation has helped to distinguish different kinds of economic activities and the level of their inter-dependence in the area.

Despite the fact that the indigenous (traditional) food processing industry in the area has recorded some successes, especially considering the fact that it (till date) handles more than 70% of the processed food, it is still not receiving the necessary attention from the government and the modern food processing bodies. It is in line with this that the paper advocates for increased research and development activities in local sourcing of food processing equipment not just in the Esan area, but other rural agricultural-based communities. This is considered very crucial; particularly for the development of small-scale and indigenous entrepreneurship in the food industry.

It is therefore in line with the above that there is the urgent need for modern food processors to collaborate with the indigenous (traditional) food processing industries scattered all over the country to achieve high quantity and quality. This is because, food processing involves different unit operation with varied equipment to carry out this operation. Unfortunately, the rural farmers/producers who are burdened with the supplying the larger population do not have access to these.

A. Appendix

Locally fabricated machines for gari production

Source: https://www.cassavaprocessing.com/Blog/locally_fabricated_garri_machine_in_nigeria_103.html

Pictures Explanation from the Left to Right.

Picture 1: The Frying Process by a woman.

Picture 2: Modern Advanced Frying Machine.

Picture 3: Modern Fabricated Dewatering Machine.

Picture 4: Locally Fabricated Dewatering Machine.

Photo of GarrI

DOI: http://dx.doi.org/10.5772/intechopen.109833

Photo of Akpu/Fufu

Photo of Elubo/Lafun in its dried form

Photo of Elubo in Powdered form

Photo of Bobozi/Tapioka/Abacha

Author details

Emmanuel Osewe Akubor[1]*, Beatrice Amili Akubor[2]
and Funmilayo Evaristus Awoyera[3]

1 Department of History, Obafemi Awolowo University, Ile-Ife, Nigeria

2 Department of Crop Protection, Obafemi Awolowo University, Ife, Nigeria

3 Department of History and International Studies, Bowen University, Iwo, Osun Sate, Nigeria

*Address all correspondence to: oseweakubor@gmail.com; akuboreo@oauife.edu.ng

References

[1] Adeniji AA, Ega LA, Akoroda MO, Adeniyi AA, Ugwu BO, de Balogun A. Cassava Development in Nigeria. Department of Agriculture, Federal Ministry of Agriculture and Natural Resources. Abuja, Nigeria: A Publication of the Federal Ministry of Agriculture and Natural Resources; 2005. [Accessed 22 September 2013]

[2] Akubor BA, Akubor EO. Women, agriculture and the development of palm oil trade in southern Nigeria. In: Singh V, editor. Interrogating Women Empowerment: The Global Experience. India: Satyam Law Publishers; 2018

[3] Ukaogo V, Akubor EO. A Diminishing Past-a Rescued Future: Essays on the Peoples, Traditions and Culture of the Esan of Southern Nigeria. New Jersey: Goldline and Jacobs; 2013

[4] Akubor EO. Colonial Domination and Cash Crop Production in Esan Division of Benin Province of Colonial Nigeria 1906-1960: A History of Agricultural Stagnation and Decline Unpublished PhD Thesis. Zaria: Department of History, Ahmadu Bello University; 2010

[5] Akubor EO. Colonialism and its impact on Food Security in Africa: A Case Study of Nigeria. In: MANDYENG: Journal of Central Nigeria Studies. A publication of the Department of History and International Studies. Nigeria: University of Jos; 2010b

[6] Akubor EO. The travail of women in crises in the niger delta area. In: The Anthropologist (Anth). Vol. 13. India: Scientific and Research International Journal, (International Bibliography of the Social Sciences); 2010c No. 1

[7] Akubor EO, Akubor BA. Gender and the Struggle against Environmental Degradation and Economic Deprivation in Africa: Niger Delta Area in Focus; Brukenthalia: Romanian Cultural History Review. Romania: University of Sibiu and Brukenthal National Museum; 2020. pp. 1102-1116

[8] Akubor EO, Akubor BA. Establishing the nexus between climate change, forced migration and food production/security in Nigeria: Research observations of historian and agriculturist. Khazanah Sosial Journal, Universitas Islam Negeri Sunan Gunung Djati, West Java. 2021;**3**(1):33-44

[9] Henn JK, Feeding the cities and feeding the peasants: What role for Africa's Farmers? World Development, II. 1983;(12)

[10] Uwaka CT, Uwaegbute AC. The role and contribution of rural women to agricultural development in eastern Nigeria. Journal of Education and Development. 1982;**2**(2):12-15

[11] Akubor EO, Eboiyehi FA. Interrogating the relevance of cultural practices/relations and linguistic evidence in reconstructing history: A case study of the people of oto esan and their neighbhor. Journal of the Institute of Culture Studies (IJICS), Obafemi Awolowo University, Ile-Ife, Vol. 14, 2021;**2021**:327-349

[12] Okoduwa AI. Time and space in Esan history: The concept and values of a people examined, Ukaogo V and Akubor, E.O (ed), A Diminshing Past–A Rescued Future: Essays on the Peoples, Traditions and Culture of Esan of Southern Nigeria, Goldline and Jacobs, New Jersey; 2013

[13] Akubor EO. Colonial Domination and Cash Crop Production in Esan Division

of Benin Province of Colonial Nigeria 1906-1960: A History of Agricultural Stagnation and Decline Unpublished PhD Thesis, Department of History, Ahmadu Bello University, Zaria; 2010;**2010**

[14] Akubor EO. 'Man and Environment: A Case Study of Esanland in Edo State of Nigeria', AAU: African Studies Review: A Journal of the Department of History and International Studies. Akungba-Akoko: Adekunle Ajasin University; 2009

[15] Iyasere E. A look at cassava production in Nigeria. International Journal of Agricultural Sciences. 2014;**5**(5):818-819. July, 2015. Available online at www.internationalscholarsjournals.org International Scholars Journals

[16] Akubor BA. Effects of Selected Legumes and Row Arrangement on Early Establishment of Cassava Seedlings in an Intercrop"; Thesis submitted in Partial Fulfilment of the requirements for the award of the Degree of Master of Science (M.Sc.) in Plant Science. Ile-Ife: Department of Crop Production and Protection, Faculty of Agriculture, Obafemi Awolowo University; 2018

[17] Oyewole O. Cassava Processing in Africa. Abeokuta Nigeria: Federal University of Agriculture; 2019 https://funaab.edu.ng/cassava-processing-in-africa/

[18] Akubor EO. The Impact of Colonialism in Nigeria: A Case Study of Esanland 1897-1960 M.A. Thesis. Zaria: Ahmadu Bello University; 2005

[19] Akubor EO, Akubor BA. Women, society and development in Nigeria: Interrogating the causes, courses and clogs. Proceedings of The Gender in the Post-2015 Development Agenda. Centre for Gender and Social Policy Studies, Obafemi Awolowo University, Ile-Ife. 2015

[20] FAO. Food Outlook - Biannual Report on Global Food Markets – November 2018: 104. License: CC BY-NC-SA 3.0 IGO. 2018 http://www.fao.org/3/ca2320en/CA2320EN.pdf

[21] FAOSTAT F. Food and Agriculture Data. 2019 http://www.fao.org/faostat/en/#data/.

[22] Denton FT, Azogu II, Ukoll MK. Cassava Based Recipes for House Hold Utilization and Income Generation. Abuja, Nigeria: AIDU, Federal Department of Agriculture; 2004

[23] Umoh VJ. Talk on food processing in Nigeria sub-sector. In: Presented at Food Processing Conference. Zaria: Ahmadu Bello University; 2012

[24] Eboyehi. Interaction and Interview with Dr. Friday Eboyehi, on Gari Production in Esan Land. Dr Eboyehi Is a Native off Irrua and a Well-Known Sociologist. He Is Currently with Center for Gender and Social Policy Studies. Ile-Ife, Osun State: Obafemi Awolowo University; 2022

[25] Akubor EO. The significance of trade and craft in social relations in Nigeria: The Esan, Benin, Igbo and Igala examples. Zaria Journal of Liberal Arts (ZAJOLA). Zaria: Ahmadu Bello University; 2007;**1**(2):22-33

[26] Okojie CG. Esan, Native Law and Custom. Yaba Lagos; 1960

Chapter 3

Breeding Cassava for End-User Needs

Ruth Naa Ashiokai Prempeh, Victor Acheampong Amankwaah, Allen Oppong and Marian Dorcas Quain

Abstract

A lot of research initiatives have gone into the breeding of cassava which has led to the development and release of over 30 cassava varieties in Ghana, of which adoption rate is 40%. This low adoption is due to inadequate promotion of improved varieties and the fact that some of the varieties do not meet end-user needs. With cassava becoming an important cash crop, it is important that breeding programmes refocus to define the market segments and objectives to facilitate the improvement of target traits such as poundability, dry matter content, starch and carotenoids that will lead to the development of varieties tailored towards end-user needs. This will in the long run promote food and nutritional security especially in low- and middle-income countries where the crop is a major staple. In addition, there should be more investment in high-throughput phenotyping to enhance the assessment and evaluation for the development of varieties with end-user traits. Subsequently, the cassava seed system should be formalized to enhance the production and dissemination of high-quality improved cassava varieties with end-user traits.

Keywords: end-user, value chain, market segments, seed system, food security

1. Introduction

Cassava (*Manihot esculenta*) is the fourth most important crop after rice, wheat and maize [1]. It is an important root crop serving as food for approximately 500 million people worldwide [2]. The crop's attributes including carbohydrate richness, availability throughout the year, tolerance to low soil fertility, and resistance to drought, pest and disease make it an attractive crop, especially to small holder farmers [3]. In addition, it is used as source of carbohydrate in animal feed. It also serves as a raw material in the manufacture of processed food and industrial products.

The world's production of cassava was reported as 278 million metric tonnes, of which Africa's contribution was about 61%, followed by Asia with 29.5%, and the Americas with 8.9% [4]. The largest producer of cassava is Nigeria (60 MT), followed by DR Congo (41.01 MT), Ghana (21.81 MT), Angola (8.78 MT) and Mozambique (5.4 MT) [4].

Cassava improvement and production are faced with many challenges. These challenges include high incidence of pests and diseases, scarcity of quality planting materials, lack of adequate technical knowledge in the use of improved technologies, inadequate modern processing equipment for end-user needs and marketing [5]. In developing countries, agricultural productivity is also constrained by limited access to improved varieties as well as biotic and abiotic stresses [6]. In addition to these constraints, seed systems for root and tuber crops including cassava have received less attention. This is usually manifested in seed multiplication and distribution which leads to challenges of low multiplication ratios from one generation to the next, bulkiness, perishability, and pest/pathogen accumulation in planting material. These issues create impediments in the establishment of commercially sustainable seed systems for the distribution of certified clean planting materials of varieties with end-user preference [7].

Many breeders have ignored the importance of breeding to meet the needs of the end-user. A lot of efforts have gone into cassava breeding in terms of agronomic traits at the neglect of breeding for the enhancement of crop palatability. Palatability of cassava (sensory characteristics) relates to end-user acceptance to ensure that improved cassava varieties are adopted and consumed by the local population in relation to the purpose for which it was released [8]. Factors such as cooking quality, size of root, appearance, odour, texture, and taste of cassava products relates to end-user consumer acceptance or preference. To meet end-user needs, breeders should take into consideration the aforementioned traits in their breeding objectives. This will facilitate breeding for varieties tailored towards end-user needs.

The success and failures of farmers to adopt newly released varieties inform breeders on their breeding objectives taking into consideration socioeconomic, gender and marketing issues. In determining where the resource investment in cassava breeding should be channelled, there is the need to pay more attention to breeding for varieties that meet the needs of different market segments. To facilitate this, the cassava breeding program in Ghana must be restructured emphasizing on target product profiles that will enhance the palatability of cassava.

2. Cassava as food security crop

Food security is when food exists for all people, at all times, and there is physical and economic access to sufficient, safe, and nutritious food. This food should meet dietary needs and food preferences for an active and healthy life [9]. Population growth and consequent increase in global consumption has called for rising demand for food globally. Additionally, the effects of climate change have enormous potential of disrupting food security [10]. These developments have generated a lot of discussions among scientists and policy makers on the adoption of approaches to ensure food security [11]. In sub-Saharan Africa (SSA), cassava is a very important food security crop for millions of people. There is therefore the need to increase breeding activities for production and improvement of its quality attributes [12]. In relation to the significant contribution of cassava to the livelihoods of African farmers and its potential of transforming the economies of Africa, cassava is reported to be amongst the six commodities defined by African Heads of State as a strategic crop for the continent [13]. Its numerous attributes make it a food security crop. Some of these attributes include tolerance to poor soils, ability to grow on marginal soils, and harvesting all year round [10].

3. Cassava breeding progress and activities

In the 1970s, most cassava breeding programs started [14], however, in the case of Ghana it started in the 1980s. In recent times, cassava has migrated from being a basic foodstuff to a cash crop serving as source of income and employment for rural populations in Africa. This is due to its potential for processing into many different products such as ethanol, starch, and high-quality cassava flour (HQCF). Crop improvement has the potential of contributing significantly to cassava transformation through the development of varieties responsive to changing needs. This could be achieved using improved technologies to support demand and supply. There is also the need for sustainable development as increases in production area expansion alone are not sustainable [15].

Cassava mosaic disease (CMD) is very devastating and was first reported in Ghana in the 1930's and government of Ghana (GoG) at that time intervened [16, 17]. Government's initiative targeted developing improved varieties tolerant to CMD to salvage cassava production in Ghana. This GoG initiative was highly welcomed since CMD is highly devastative severely affecting all existing local accessions at early stages and therefore meriting a lot of attention [18]. Superior varieties were introduced from sister countries in Africa and the Caribbean. This was followed by making several crosses between local accessions and the introduced superior clones followed by selection of desirable traits including CMD tolerance. The varieties were named as Queen, Gari, Williams and Ankrah and these were subsequently released in 1935 [17]. From 1993 to date, many improved cassava varieties which are tolerant to CMD and are high yielding have been released by the Council for Scientific and Industrial Research Institute (CSIR), the universities and other research institutes [19]. Aside these progress, use of molecular markers were introduced for the selection of desirable traits with emphasis on CMD tolerance. Approximately all the varieties released in Ghana are white pulp. To meet demands of malnutrition and malnourished populace, cassava breeding programmes targeted developing yellow-fleshed varieties in the last decade. In addition to the 25 released cassava varieties in Ghana, nine yellow-fleshed varieties have been released to meet consumer needs for gari production and other products.

3.1 Conventional breeding

Conventional planting breeding methods have enormous benefits such as high rates of return among the investments in agricultural research. In this regard, cassava breeding has achieved enormous benefit from technological input. New varieties released in Africa, Asia and Latin America with conventional breeding efforts have to some extent met the needs of farmers, processors, and consumers for income generation. Using participatory breeding approach, plant breeding is advantageous for subsistence farming where many subtle criteria define the success of a given variety and subsequent chance of adoption by farmers [20].

The same conventional cassava breeding method are used by many breeding programs, with few variations. This entails producing full- or half-sib families in crossing blocks. Cassava is highly heterozygous, hence progenitors generated from crosses are genetically diverse. F_1 seedlings are genetically distinct therefore production of enough cuttings for multilocation trials takes several years. In addition, the multiplication rate is usually low (1:10) [21]. Due to its long breeding cycle, which is approximately 12 months, the development of improved varieties is time consuming [12].

3.2 Biotechnological interventions for cassava improvement

For a constantly changing world, plant biotechnology provides a wide range of opportunities that can facilitate cassava production [22]. Great advances have been made in cassava using biotechnology in the past 30 years [22]. This section will focus on transgenics, gene editing and marker-assisted techniques.

3.2.1 Transgenics

Genetic engineering advances can significantly speed up the development of improved varieties with enhanced yield, nutritional quality, increased disease and pest resistance, as well as improved starch yield and quality [23]. Improvement of cassava using transgenics help in overcoming crossability barriers and facilitates the development of improved varieties with end-user traits. It however has some challenges such as low level of awareness, lack of appropriate facilities and infrastructure [24].

Cassava genetic transformation efforts have been used in improving traits such as CMD resistance, increase in protein content from 40 to 130% and improved starch content and quality [25]. For the development of transgenic cultivars, desirable genes are cloned, vectors are constructed, crops are transformed, and subsequently, screening and identification of transformed lines are conducted [23, 26]. Through a concerted effort of several laboratories for about 25 years, transformation of cassava and the use of *Agrobacterium tumefaciens* or particle bombardment as gene delivery system became a reality. It has been possible to obtain transgenic plants of cassava that expresses marker and selectable genes of agronomic traits.

3.2.2 Gene editing

Improvement of cassava is of utmost importance since it is the fourth most important crop and uniquely occupies an important position. Technologies of gene editing based on zinc-finger nucleases (ZFNs), transcription activator-like effector nucleases (TALENS), and clustered regularly interspaced short palindromic repeats (CRISPR) using engineered nucleases induces double strand breaks (DSB) at known DNA sequence within the genome. Repair at the target site subsequently, introduces variation via error prone non-homologous end joining (NHEJ) [27]. CRISPR-associated protein 9 (Cas9) have been shown to improve target traits to meet end-user needs [28]. This technology has been used for the modification of phytoene desaturase [27], increment in carotenoid content and production of waxy starch [29] in cassava.

Gene editing enables precision breeding, reduces time of breeding cycle with fewer plant generations and help in the development of varieties with enhanced nutrition [30]. In crops where there is little to no information on genome sequence, it will be impossible to identify potential targets of interest for editing [31, 32] and subsequently impossible to develop new improved varieties using this technology.

3.2.3 Marker-assisted breeding

Genetic diversity is of paramount importance in crop improvement. Different molecular markers have been used to assess the genetic diversity among crop varieties. Genetic variation from some African countries have been assessed using molecular markers such as simple sequence repeats (SSRs) and single nucleotide polymorphism markers (SNPs) [33, 34]. Advanced technologies such as genotyping-by-sequencing (GBS) have also been used for the identification of varieties in Nigeria [35] and Ghana [6]. For enhancement of conventional plant breeding, high density genome-wide markers, combined with statistical tools have been used for the identification and validation of trait markers for marker-assisted selection (MAS) and implementation of genomic selection. In addition, whole genome sequencing has also been used for the discovery of markers in cassava breeding populations. Identification of quantitative trait loci (QTLs) in breeding populations have been achieved using high-density markers in association and linkage mapping. These detected QTLs provide detailed information for designing diagnostic markers for MAS.

The use of molecular markers greatly increases the efficiency and effectiveness of breeding. It can be carried out in the seedling stage and reduces the breeding cycle. Some limitations of the technique are high startup expenses, recombination between the marker and the gene of interest may occur leading to false positives, non-transferability of markers from one population to other populations and imprecise estimates of QTL may result in slower progress than expected [36].

4. Released cassava varieties, markets, and adoption

Different cassava varieties have been developed and released by CSIR-Crops Research Institute (CRI), Universities in Ghana and other institutes targeting different market segments. These varieties have been adopted for use and processed into different products to meet end-user needs.

4.1 Released cassava varieties

Over 30 cassava varieties with varying uses have been released in Ghana (**Figure 1**). **Table 1** shows released varieties, quality attributes and their uses.

Figure 1.
Vegetative phase (a) and fresh roots (b) of some released cassava varieties in Ghana.

Variety	Dry matter content (%)	Other quality attributes	Uses
Afisiafi	32	Not poundable	Starch, flour and gari
Abasafitaa	35	Not poundable	Starch, flour and gari
Tek-bankye	30	Poundable	Fufu, gari and "ampesi"
Nyeri-kogba	33	Not Poundable in the dry season	Starch, gari, flour and "tuo zaafi"
Eskamaye	33	Not poundable in the dry season	Starch, flour, gari and "tuo zaafi"
Fil-Ndiakong	36	Not poundable in the dry season	Starch, flour, gari and "tuo zaafi"
Nkabom	32	Poundable	Starch and fufu
IFAD	30	Poundable	Starch and fufu
CRI-Agbelefia	—	24% Starch, not poundable	Starch and gari
CRI-Essam Bankye	—	19.8% Starch, not poundable	Starch and flour
CRI-Bankye Hemaa	—	21% Starch, poundable	Fufu and bakery products
CRI-Doku Duade	—	24% Starch, not poundable	Starch
Capevars Bankye	—	Starch >25%, poundable	Starch, fufu, "ampesi", gari, flour, "agbelima"
Bankye Botan	—	Poundable in the dry season	Starch, gari, "agbelima", "konkonte"
CRI-Ampong	36	Poundable	Starch, flour and bakery products
CRI-Broni Bankye	33	Not poundable	Starch, flour and bakery products
CRI-Otuhia	39	Not poundable	Starch and flour
CRI-Duade Kpakpa	37	Poundable	Starch, flour, fufu and industrial alcohol
CRI-Amansan Bankye	38	Not poundable	Flour and bakery products
CRI-AGRA Bankye	32	Not poundable	Starch and flour
CRI-Dudzi	38	Not poundable	Starch and flour
CRI-Abrabopa	40	Not poundable	Starch
CRI-Lamesese	39	Poundable, pro-vitamin A	Flour and fufu
CRI-Bediako	33	Poundable	Starch, flour and fufu
CRI-Bankye	30	Poundable	Starch, flour and fufu
Nyonku Agbeli	26	CP: 36.75 μg/g, pro-vitamin A: 9.6 μg^{-1}	Flour and gari
Kpomu Agbeli	26	CP: 32.58 μg/g, pro-vitamin A: 7.97 μg^{-1}	Flour and gari
Tetteh Bankye	21	Poundable, CP: 31.91 μg/g, pro-vitamin A: 6.91 μg^{-1}	Flour, gari and fufu
Fufuohene Bankye	41	Poundable, CP: 32.80 μg/g	Starch, flour, gari and fufu
Ampesi Hemaa Bankye	40	Poundable, CP: 34.71 μg/g	Starch, flour, fufu and "ampesi"

Variety	Dry matter content (%)	Other quality attributes	Uses
CRI-Kent	30	CP: 48.15 mg/100 g, TCC: 7.8 μg/g	Gari, Chips
CRI-Kyerewaa	33	Poundable, CP: 42.03 mg/100 g, TCC: 4.12 μg/g	Gari, chips
CRI-Peprah	30	CP: 40.38 mg/100 g, TCC: 12.70 μg/g	Gari, chips
CRI-Manu	30	CP: 63.11 mg/100 g, TCC: 10.40 μg/g	Gari, chips

CP, cyanogenic potential; TCC, total carotenoid content [19].

Table 1.
Quality attributes and uses of released cassava varieties.

4.2 Potential markets

Cassava could be marketed fresh for cooking or could be processed into many different products of industrial and economic value. Some of these products include gari, starch, ethanol, high-quality cassava flour (HQCF), and other products. For example, in Nigeria, the cassava commercialization and market promotion programme were implemented between 2002 and 2008. Processed products such as HQCF, starch, glucose syrup and ethanol were produced. To replace some imported raw materials with the intermediate cassava products, this initiative was backed with policy. The programme increased production enormously by 10 million tonnes in the six years of implementation [37] and Nigeria is still reaping a lot of benefits from this initiative.

In Ghana, the industrial starch production was promoted by GoG in 2001. This was spearheaded by the Cooperative Village Enterprise (COVE) program. The initiative resulted in the establishment of a starch processing plant. The presence of this processing plant facilitated the buying of roots from villages in that vicinity leading to job creation for most of the citizenry. Though some challenges were encountered, there was increase in cassava production and processing at that time [37]. Currently, there are new business opportunities for cassava processing where private entrepreneurs have shown a lot of interest in products such as ethanol, starch, HQCF, chips for animal feed. To meet the needs of these different markets, the cassava breeding programme must redefine and confirm the country's cassava market segments and develop target product profiles.

4.3 Adoption of released varieties

A lot of research has gone into the development and release of new varieties due to the potential of cassava as a food security and industrial crop, as well as other benefits such as its multiple opportunities for poverty reduction and nourishment for vulnerable population. The adoption rate of improved cassava varieties in Ghana have been reported to be 40% [17]. This low adoption of improved varieties could be attributed to inadequate promotion and the fact that some varieties do not meet end-user needs, hence most farmers still rely on landraces [38]. Consequently, there is the need to change this trend through a lot of promotion of the good attributes of improved varieties for the attraction of both local and international markets. In addition, the breeding programmes in Ghana need to change their breeding strategy to develop new varieties to meet end-user needs.

5. Target traits for cassava improvement to meet end-user needs

Cassava has a lot of potentials to tackle malnutrition in addition to generation of income especially in parts of the world where food and nutrition security is a major challenge. A lot of breeding work has concentrated on agronomic traits with relatively less emphasis on nutritional quality and end-user traits. To make more progress to breed for varieties that will meet end-user needs, it is important to pay more attention to some added traits such as improvement in starch content and quality, dry matter content, mealiness/poundability in addition to the usual agronomic traits. Other traits such as early bulking is also important especially in this era of climate change.

For the acceptance of cassava for value chain actors, dry matter content (DMC) is an important character for variety acceptance. DMC is referred to as true biological yield or economic yield which is controlled by many genes [39, 40]. Majority of cassava accessions have DMC between 20 and 40% [39]. DMC accumulation is influenced by genetic and environmental factors. These factors include age of the crop, efficiency of canopy to trap sunlight, season, and location effects [41]. Therefore, in breeding for high DMC, there is the need to consider these factors.

Cassava starch is the cheapest and the most preferred because of its many positive characteristics such as high paste clarity, relatively good stability to retrogradation and swelling capacity. Other qualities include low protein complex, and good texture [41]. Most cassava varieties released in Ghana have starch content between 20 and 30%. As Ghana become more industrialized with the one district one factory and planting for food and jobs initiatives, there will be the need to breed for new varieties with higher starch content than existing varieties currently available. In addition, the breeding program should have replacement strategies to replace cultivated varieties that have declined in starch content. This will result in the release of superior varieties with high starch content to meet the demands of the starch growing industry in Ghana and beyond.

Cassava is boiled and eaten as "ampesi" or pounded together with plantain to make fufu, a delicacy for most people in Ghana. About 70% of cassava produced is used for fufu, making poundability a major end-user trait. Out of over 30 improved varieties released, about 60% are poundable, however most of them are not poundable all year round. It is imperative for breeders to pay more attention to poundability as a major target trait in the breeding programme. Another important organoleptic attribute of poundability in recent times is the texture of boiled cassava [42]. This is crucial for the release of end-user preferred varieties as these are principal quality attributes of boiled cassava roots [43].

The process of increasing micronutrient amount in crops through plant breeding, transgenic techniques, or agronomic practices is termed biofortification. This is a feasible means of reaching rural populations with limited access to diverse diets or other micronutrient intervention [44]. More than two billion individuals are affected by micronutrient deficiencies [45]. There should therefore be the need to breed for more biofortified cassava varieties to feed the vulnerable populace. In addition, there should be a balance between the potential benefit of increasing market demand for biofortified cassava, thereby making it more attractive for farmers to grow [44]. Most cassava varieties have white to cream root pulp. The colour of the pulp is closely linked to carotenoid content. Development of high carotenoid biofortified cassava varieties have been initiated by HarvestPlus [46] and through the project, nine yellow-fleshed varieties with good carotenoid content have been released in Ghana to help meet the needs of the vulnerable population with vitamin A deficiency.

6. Phenotyping tools for breeding for end-user preference

For conventional breeding, investment, and use of new tools for phenotyping should be taken seriously especially in SSA. This will enhance modernization of the breeding programme making it more attractive to the youth especially in this era of climate change. To facilitate genetic studies and improve genetic gains, accurate phenotyping is key. However, getting accurate phenotypic data for breeding to meet end-user needs remains a challenge [12]. Conventional use of colour intensity in the measurement of carotenoid content in cassava roots is sometimes very difficult [46]. Near-infrared spectroscopy (NIRS) has been applied in some countries for the prediction of nutritional constituents of cassava. This holds a lot of promise in exploiting carotenoid levels and other quality traits. In addition, the use of high-performance liquid chromatography (HPLC) would complement high-throughput phenotyping efforts in breeding for varieties tailored towards end-user needs. The breeding programme in Ghana must invest in high-throughput phenotyping equipment for assessing micronutrient levels and other quality traits.

7. Cassava seed systems for dissemination of end-user preferred varieties to stakeholders

Conventional system of propagation using stems has disadvantages such as low multiplication rate (7–10 new stakes/mature plant/cycle) and accumulation of pathogens with time. This system also takes long to get enough planting materials (8–14 months) [22]. Due to these challenges, complementing the efforts of conventional multiplication with innovative rapid propagation methods such as tissue culture and semi-autotrophic hydroponics (SAH) hold a lot of prospects [47], reported that a major challenge of cassava breeding programmes compared with private seed business is that the former lack efficient seed systems to produce and distribute planting materials of newly improved varieties with end-user needs. Putting in place a resilient and formal seed system starting from the tissue culture and molecular diagnostic laboratories, as well as screen house facility is key for increasing the production of varieties with end-user traits. In the case of Ghana, there are facilities for cleaning, indexing and mass production of improved cassava varieties. However, the cassava seed system is not well structured, hence this must be improved to enhance the dissemination of improved cassava varieties with end-user traits.

8. Linkage between breeding and market

Well targeted research initiatives are required for the initiation of commercial operation of cassava value chain. Value chain commercialization, linking breeding activities to the market, requires relevant technologies and market innovations to be tested at pilot scale. This will facilitate lessons learnt from technology adaptation, the challenges and other precondition for success. To link breeding and market, this calls for research and development initiatives with the primary objective of providing solutions to constraints militating against the efficient operation of the cassava value chain [37].

9. Future prospects and way forward

Cassava breeding activities keep on revolutionizing. To meet end-user needs and the growing demand for more cassava production, traits targeted should be chosen carefully. The Ghana cassava breeding programmes should redefine the country's cassava market segments, change the breeding strategy to develop new varieties with traits such as poundability, DMC, starch and carotenoids to meet end-user needs. They must also invest in high-throughput phenotyping equipment to enhance the development of varieties with end-user traits. The breeding programme should also look at formalizing the cassava seed system to enhance the production and dissemination of high-quality improved cassava varieties with end-user traits.

10. Conclusions

A lot of efforts have gone into cassava breeding in terms of agronomic traits. Consequently, most varieties do not meet end-user needs. This has resulted in relatively low adoption and most farmers still cultivate landraces. To overcome these challenges, the cassava breeding programmes in Ghana should identify the different market segments, develop target product profiles for the different market segments, develop improved varieties with end-user traits, outdoor and promote varieties, formalize the cassava seed system for the production and dissemination of clean planting materials. When all these are fulfilled, improved varieties released will be greatly adopted and utilized by farmers.

Acknowledgements

The authors wish to express their sincere thanks to scientists and technicians at the CSIR-Crops Research Biotechnology Laboratory and Cassava Breeding Programme who contributed in one way or the other towards the preparation of this book chapter.

Conflict of interest

The authors declare no conflict of interest.

Author details

Ruth Naa Ashiokai Prempeh*, Victor Acheampong Amankwaah, Allen Oppong and Marian Dorcas Quain
CSIR-Crops Research Institute, Kumasi, Ghana

*Address all correspondence to: ginathompsongh@yahoo.com

References

[1] FAO and IFAD. The world cassava economy: Facts, Trends and Outlook. Vol. 59, International Fund for Agricultural Development. Rome, Italy: Food & Agriculture Organization; 2000

[2] Balagopalan C. Cassava utilization in food, feed and industry. In: Hillocks RJ, Thresh JM, Bellotti AC, editors. Cassava: Biology, Production and Utilization. New Delhi, India: CAB International; 2001. pp. 301-318

[3] Tesfaye T, Atnafua B, Engida T, Getachew WM, Tewodros M, Wubshet B, et al. Performance of cassava (*Manihot esculanta*. Cratz) clones in potential and low moisture stressed areas of Ethiopia. African Journal of Agricultural Research. 2017;**12**(20):1738-1746

[4] FAOSTAT. FAO Statistics, Food and Agriculture Organization of the United Nations [Internet]. 2020. Available from: http://www.fao.org/faostat/ [Cited: 28 November 2022]

[5] Akinnagbe O. Constraints and strategies towards improving cassava production and processing in Enugu North Agricultural Zone of Enugu State, Nigeria. Bangladesh Journal of Agricultural Research. 1970;**35**(3):387-394

[6] Rabbi IY, Kulakow PA, Manu-Aduening JA, Dankyi AA, Asibuo JY, Parkes EY, et al. Tracking crop varieties using genotyping-by-sequencing markers: A case study using cassava (*Manihot esculenta* Crantz). BMC Genetics. 2015;**16**(1):1-11. DOI: 10.1186/s12863-015-0273-1

[7] Almekinders CJM, Walsh S, Jacobsen KS, Andrade-Piedra JL, McEwan MA, de Haan S, et al. Why interventions in the seed systems of roots, tubers and bananas crops do not reach their full potential. Food Security. 2019;**11**(1):23-42

[8] Bechoff A, Tomlins K, Fliedel G, Becerra Lopez-lavalle LA, Westby A, Hershey C, et al. Cassava traits and end-user preference: Relating traits to consumer liking, sensory perception, and genetics. Critical Reviews in Food Science and Nutrition. 2018;**58**(4):547-567. DOI: 10.1080/10408398.2016.1202888

[9] Pinstrup-Anderson P. Food security: Definition and measurement. Food Security. 2009;**1**:5-7

[10] Reincke K, Vilvert E, Fasse A, Graef F, Sieber S, Lana MA. Key factors influencing food security of smallholder farmers in Tanzania and the role of cassava as a strategic crop. Food Security. 2018;**10**(4):911-924

[11] Brüssow K, Faße A, Grote U. Implications of climate-smart strategy adoption by farm households for food security in Tanzania. Food Security. 2017;**9**(6):1203-1218

[12] Mbanjo EG, Rabbi IY, Ferguson ME, Kayondo SI, Eng NH, Tripathi L, et al. Technological innovations for improving cassava production in sub-Saharan Africa. Frontiers in genetics. 21 Jan 2021;**11**:1829

[13] Feleke FB, Berhe M, Gebru G, Hoag D. Determinants of adaptation choices to climate change by sheep and goat farmers in Northern Ethiopia: The case of Southern and Central Tigray, Ethiopia. Springerplus. Dec 2016;**5**:1-5

[14] Ceballos H, Pérez JC, Barandica OJ, Lenis JI, Morante N, Calle F, et al. Cassava breeding I: The value of breeding

value. Frontiers in Plant Science. 2016;**7**:1-12

[15] Dixon AGO, Ssemakula G. Prospects for cassava breeding in Sub-Saharan Africa in the next decade. Journal of Food, Agriculture and Environment. 2008;**6**(3-4):256-262

[16] Doku EV. Root crops in Ghana. Ghana Journal of Science. 1966;**6**:15-36

[17] Acheampong PP, Danquah EO, Agyemang K, Dankwa KO. Research and development for improved cassava varieties in Ghana: Farmers' adoption and effects on livelihoods. In: Frediansyah A, editor. Cassava-Biology, Production and Use. London, UK: IntechOpen; 2021. p. 13. Available from: https://www.intechopen.com/books/advanced-biometric-technologies/liveness-detection-in-biometrics

[18] Wong TM. The current pandemic of cassava mosaic virus disease in Uganda. In: Jones DG, editor. The Epidemiology of Plant Diseases. Dordecht: Springer; 1998. pp. 423-443

[19] MoFA. Catalogue of crop varieties released and registered in Ghana, Accra. 2019

[20] Mulualem T, Bekeko Z. Assessment of conventional breeding on cassava and its physiological adaptive mechanisms: Implication for moisture stress. Asian Journal of Agricultural Research. 2015;**9**(2):38-54

[21] Ceballos H, Rojanaridpiched C, Phumichai C, Becerra LA, Kittipadakul P, Iglesias C, et al. Excellence in cassava breeding: Perspectives for the future. Crop Breeding, Genetics and Genomics. 3 Apr 2020;**2**(2):1-31

[22] Chavarriaga-Aguirre P, Brand A, Medina A, Prías M, Escobar R, Martinez J, et al. The potential of using biotechnology to improve cassava: A review. In Vitro Cellular & Developmental Biology. Plant. 2016;**52**(5):461-478. DOI: 10.1007/s11627-016-9776-3

[23] Kamthan A, Chaudhuri A, Kamthan M, Datta A. Genetically modified (GM) crops: Milestones and new advances in crop improvement. Theoretical and Applied Genetics. 2016;**129**(9):1639-1655

[24] Pérez-Massot E, Banakar R, Gómez-Galera S, Zorrilla-López U, Sanahuja G, Arjó G, et al. The contribution of transgenic plants to better health through improved nutrition: Opportunities and constraints. Genes & Nutrition. 2013;**8**(1):29-41

[25] Liu J, Zheng Q, Ma Q, Gadidasu KK, Zhang P. Cassava genetic transformation and its application in breeding. Journal of Integrative Plant Biology. 2011;**53**(7):552-569

[26] Yu H, Yang Q, Fu F, Li W. Three strategies of transgenic manipulation for crop improvement. Frontiers in Plant Science. 2022;**13**(July):1-15

[27] Odipio J, Alicai T, Ingelbrecht I, Nusinow DA, Bart R, Taylor NJ. Efficient CRISPR/cas9 genome editing of phytoene desaturase in cassava. Frontiers in Plant Science. 2017;**8**:1-11

[28] Fondong VN, Rey C. Recent biotechnological advances in the improvement of cassava. In: Waisundara V, editor. Cassava. London, UK: IntechOpen; 2018. p. 13. Available from: https://www.intechopen.com/books/advanced-biometric-technologies/liveness-detection-in-biometrics

[29] Bull SE, Seung D, Chanez C, Mehta D, Kuon JE, Truernit E, et al.

Accelerated ex situ breeding of GBSS-and PTST1-edited cassava for modified starch. Science Advances. 5 Sept 2018;**4**(9):eaat6086

[30] Zaruk D. 20 Benefits from Agricultural Genome Editing. 2020. [Cited 2022 Dec 15]. Available from: https://seedworld.com/20-benefits-from-agricultural-genome-editing/

[31] Hahn F, Nekrasov V. CRISPR/Cas precision: Do we need to worry about off-targeting in plants? Plant Cell Reports. 2019;**38**(4):437-441. DOI: 10.1007/s00299-018-2355-9

[32] Venezia M, Creasey Krainer KM. Current advancements and limitations of gene editing in orphan crops. Frontiers in Plant Science. 22 Sep 2021;**12**:742932

[33] Kamanda I, Blay ET, Asante IK, Danquah A, Ifie BE, Parkes E, et al. Genetic diversity of provitamin-A cassava (*Manihot esculenta* Crantz) in Sierra Leone. Genetic Resources and Crop Evolution. 2020;**67**(5):1193-1208. DOI: 10.1007/s10722-020-00905-8

[34] Adjebeng-Danquah J, Manu-Aduening J, Asante IK, Agyare RY, Gracen V, Offei SK. Genetic diversity and population structure analysis of Ghanaian and exotic cassava accessions using simple sequence repeat (SSR) markers. Heliyon. 2020;**6**(1):e03154. DOI: 10.1016/j.heliyon.2019.e03154

[35] Wossen T, Girma G, Abdoulaye T, Rabbi I, Olanrewaju A, Alene A, et al. The Cassava Monitoring Survey in Nigeria [Internet]. Available from: www.iita.org

[36] Hasan N, Choudhary S, Naaz N, Sharma N, Laskar RA. Recent advancements in molecular marker-assisted selection and applications in plant breeding programmes. Journal, Genetic Engineering & Biotechnology. 2021;**19**(1):1-26

[37] Abass A, Mlingi N, Ranaivoson R, Zulu M, Mukuka I, Abele S, et al. Potential for Commercial Production and Marketing of Cassava: Experiences from the Small-Scale Cassava Processing Project in East and Southern Africa [Internet]. Available from: www.iita.org

[38] Acheampong PP, Acheampong LD. Analysis of adoption of improved cassava (*Manihot esculenta*) varieties in Ghana: Implications for agricultural technology disseminations. International Journal of Food and Agricultural Economics. 2020;**8**(3):233-246

[39] Amelework AB, Bairu MW. Advances in genetic analysis and breeding of cassava (*Manihot esculenta* Crantz): A review. Plants. 2022;**11**(12):1-19

[40] Prasannakumari V, Gopalakrishnan Heinining Nair A, Mohan C. Identification of quantitative trait loci (QTLs) conferring dry matter content and starch content in cassava (*Manihot esculenta* Crantz). American Journal of BioScience. 2021;**9**(1):1

[41] Ayetigbo O, Latif S, Abass A, Müller J. Comparing characteristics of root, flour and starch of biofortified yellow-flesh and white-flesh cassava variants, and sustainability considerations: A review. Sustainability. 30 Aug 2018;**10**(9):3089

[42] Uchendu K, Njoku DN, Paterne A, Rabbi IY, Dzidzienyo D, Tongoona P, et al. Genome-wide association study of root mealiness and other texture-associated traits in cassava. Frontiers in Plant Science. 17 Dec 2021;**12**:770434

[43] Hongbété F, Mestres C, Akissoé HN, Brigitte P. Effects of cultivar and

harvesting conditions (age, season) on the texture and taste of boiled cassava roots. Food Chemistry. 2011;**126**(1):127-133

[44] Bouis HE, Saltzman A. Improving nutrition through biofortification: A review of evidence from HarvestPlus, 2003 through 2016. Global Food Security. 2017;**12**:49-58

[45] FAO, IFAD, WFP. The state of food insecurity in the world 2015. In: Meeting the 2015 international hunger targets: taking stock of uneven progress. Journal of Nutrition. Rome: FAO; 2015:1576-1580

[46] Ceballos H, Davrieux F, Talsma EF, Belalcazar J, Chavarriaga P, Andersson MS. Carotenoids in cassava roots. Carotenoids. 14 Jun 2017:189-221

[47] Ceballos H, Hershey C, Iglesias C, Zhang X. Fifty years of a public cassava breeding program: Evolution of breeding objectives, methods, and decision-making processes. Theoretical and Applied Genetics. 2021;**134**(8):2335-2353. DOI: 10.1007/s00122-021-03852-9

Section 2

Cassava Biotechnology and Soil Modification

Chapter 4

Plant Regeneration from Cassava Protoplasts

Wen Feng, Hai-Tian Fu, Yan-Chun Luo and Jian-Qi Huang

Abstract

Cassava is an important crop for food, feed, and industrial raw materials. Given that traditional conventional breeding is restricted by various factors, biotechnology breeding has become an important breeding method. Tissue culture regeneration is the basis of biotechnology breeding. This chapter reviews the establishment and development of cassava tissue culture and regeneration systems and the technical processes of tissue culture and regeneration starting from the induction of explants of tissue-cultured cassava plantlets to embryogenic calli, isolation to protoplasts, culture to embryogenic calli followed by differentiation into embryos, and then sprouting, stemming, and rooting into complete plants. This chapter focuses on the technical processes from protoplast to complete plant and summarizes the important influencing factors of protoplast regeneration, which is the key and difficult point in the entire regeneration process of cassava protoplasts. This chapter aims to provide technical guidance for cassava protoplast regeneration, offer useful inspiration and reference for cassava tissue culture, and lay a foundation for the genetic improvement of cassava.

Keywords: cassava, biotechnology, tissue culture, friable embryogenic callus, protoplast, somatic embryogenesis

1. Introduction

Cassava (*Manihot esculenta* Crantz), a perennial shrub of the Euphorbiaceae family, is widely cultivated in tropical and subtropical regions. It is a root crop that is a staple food for approximately 800 million people worldwide [1, 2]. It is also an important raw material for the production of starch, processed food, and biofuels [3, 4]. Its tubers, tender branches, and leaves are commonly used as animal feed [5]. Cassava also has some advantages, such as tolerance to adverse environmental conditions, adaptation to poor soils, flexible harvest times, and the capability for growth under marginal conditions [6].

Viral diseases, insect pests, toxic cyanogenic glucosides, postharvest physiological deterioration, and low root protein content roots are problems in cassava cultivation [7, 8]. Traditional conventional cassava breeding is restricted by several problems, such as high genotype heterozygosity, long life cycle, low natural fertility, poor seed set, and low seed germination rates [9–11]. Biotechnology breeding is a supplement to traditional breeding methods. With the development of molecular breeding and genetic engineering, biotechnology breeding requires an effective regeneration system [12, 13].

This chapter provides a review of the establishment and development of cassava tissue culture and regeneration systems and the tissue culture and regeneration technology of cassava starting from the induction of the explants of tissue-cultured cassava plantlets into embryogenic calli and then into protoplasts, followed by culturing into complete plants in the cyclic process of plant regeneration from cassava protoplasts. It also provides information on experiences and skills in protoplast regeneration to lay a foundation for the genetic improvement of cassava.

2. Establishment and development of cassava tissue culture and regeneration systems

Cassava tissue culture and regeneration technologies have been continuously developed since Kartha et al. cultured the apical meristem of cassava and obtained complete plants of five varieties for the first time in 1974 [14].

2.1 Organogenesis

Cassava axillary buds were cultured on medium with a high concentration of 6-benzylaminopurine (6-BA, 10 mg L^{-1}) to form a round compact bulb-like structure and then transferred onto medium with a low concentration of 6-BA (1 mg L^{-1}) for multiple shoot production; this approach was an efficient mass propagation system for cassava [15, 16].

On the basis of the establishment of the plant regeneration pathway of cassava somatic embryogenesis, cotyledons formed from primary embryos, secondary embryos, and circulating embryos could regenerate plants through organogenesis in medium containing 1.0 mg L^{-1} 6-BA and 0.5 mg L^{-1} IBA [17]. The primary embryo has low ability for cotyledon organogenesis, whereas the circulating embryo has the highest ability for cotyledon organogenesis [18]. Cassava explants for organogenesis could be derived from the axillary buds, stem tips, young leaf lobes, and cotyledons of primary, secondary, and circulating embryos.

2.2 Somatic embryogenesis

Somatic embryogenesis has been widely developed since Stamp and Hemhaw first reported in 1982 that embryoids could be successfully induced from the cotyledons and hypocotyls of cassava zygotic embryos [19]. The four-step method of the somatic embryogenesis and plant regeneration of cassava has been established as follows: (1) induction of somatic embryos on medium containing 2,4-dichlorophenoxyacetic acid (2,4-D) and other auxins; (2) maturation or germination on medium containing a low concentration of 6-BA; (3) growth on medium containing a high concentration of 6-BA and development into stems; and (4) rooting in low naphthaleneacetic acid (NAA) concentration or hormone-free medium [20–22]. In steps 1 and 2 of circulation, secondary and cyclic somatic embryos could be generated, forming a cyclic somatic embryogenesis system, and cyclic embryo explants could be induced into embryos more significantly [20, 22].

Since then, most studies on somatic embryogenesis performed optimization in accordance with different factors, such as genotype, explant type, and hormone type.

Explants have also been developed from the hypocotyls and cotyledons of the initial zygotic embryo and the apical buds, young leaf lobes, axillary buds, flower tissues, and cotyledons of primary, secondary, and cyclic somatic embryos. The development of somatic embryogenesis and plant regeneration laid a foundation for the induction of friable embryogenic callus (FEC) [23] and genetic transformation in cassava [24].

2.3 Friable embryogenic callus (FEC) induction

FEC is an important research material in genetic and cell engineering. It has been developed for plants, and zygotic embryos are usually the preferred explant materials for FEC induction [25]. The zygotic embryos of cassava have extremely heterogeneous and unclear genotypes and are therefore unsuitable as explants.

Taylor et al. described the initiation of FEC for the first time. They utilized young leaves to induce embryogenic calli [23]. After embryogenic calli were generated, high-quality embryonic tissues were continuously subcultured on Gresshoff and Doy (GD) medium containing picloram to produce a small cell cluster that was composed of dozens of cells with diameters of 1 mm; these tissues were considered as fragile embryogenic calli from which highly totipotent embryogenic suspension cultures were established [23].

Since then, many reports on FEC induction have been published [26–29]. However, considering that cassava is a gene-dependent crop, FEC cannot be induced successfully for every variety [10, 30].

Successful FEC induction has laid a foundation for genetic transformation [28, 31, 32], CRISPR/Cas9 genome editing technology [33, 34], protoplast culture and regeneration [35, 36], and somatic hybridization [37].

2.4 Protoplast culture and regeneration

2.4.1 Culture and regeneration of mesophyll protoplasts

Protoplasts were separated from the leaves of tissue-cultured seedlings and cultured in a double-layered solid and liquid medium inserted with short glass rods evenly and vertically. No glass rod was inserted in the control culture, and the remaining cultures were all the same. The protoplasts divided continuously to form visible calli only in the medium inserted with glass rods [38].

After a long time, plant regeneration from the leaf mesophyll-derived protoplasts of cassava was reported in 2022 [39]. Prior to this report, only one successful report of plant regeneration from protoplasts isolated from cassava leaves had been published [40], and other scholars could not repeat this process [37, 38].

2.4.2 Culture and regeneration of protoplasts from embryos and embryogenic calli

The protoplasts separated from secondary embryos were tested in more than 50 media to form visible calli, a small number of which formed adventitious roots but never formed adventitious buds or embryos [41].

At present, most methods for protoplast regeneration involve inducing FEC and establishing an embryogenic suspension culture for the separation of protoplasts. In the 1990s, a breakthrough was made in the research on cassava protoplast regeneration. On the basis of establishing FEC induction technology and its suspension culture system, research on the isolation, purification, and culture of cassava protoplasts from FEC was carried out, and plants were regenerated. However, regeneration efficiency was low mainly due to the low efficiency of protoplast-derived calli for differentiating into somatic embryos and the low germination efficiency of mature embryos, which is the bottleneck in cassava protoplast regeneration [35].

Subsequently, cassava protoplast regeneration was not reported for a long time. In 2012, Wen et al. improved the yield and activity of isolated protoplasts on the basis of their predecessors. Callus-derived protoplasts were first subjected to suspension culture in suspension culture medium (SH) liquid medium, and then cultured in somatic embryo emerging medium (MSN) solid medium [36]. The bottleneck mentioned by Sofiari et al. [35] was broken, and the regeneration efficiency of protoplasts was greatly improved. On the basis of the established protoplast regeneration technology system, tetraploid cassava plants were regenerated *via* protoplast electrofusion [37].

3. Plant regeneration from cassava protoplasts

In this section, cassava tissue culture and regeneration technology is mainly reviewed starting with the induction of explants of tissue-cultured cassava plantlets into embryogenic callus; followed by the isolation of protoplasts; the culture and differentiation into embryos of embryogenic calli; and sprouting, stemming, and rooting into a complete plant. This section focuses on the technical process from the isolation of protoplasts to the generation of a complete plant (**Figure 1**).

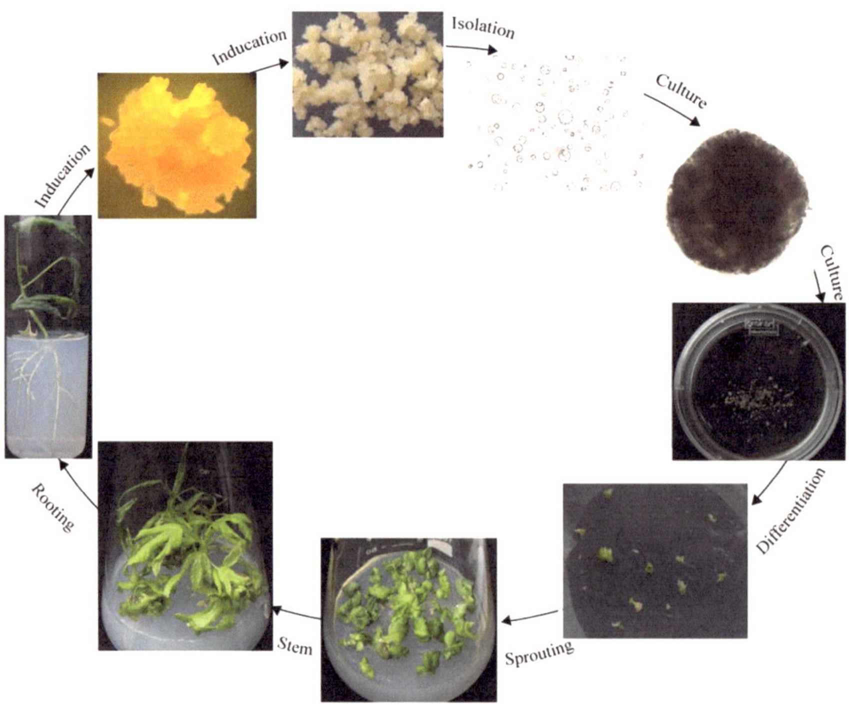

Figure 1.
Cyclic process of plant regeneration from cassava protoplasts.

3.1 Explants

Young leaf lobes of tissue-cultured cassava seedlings with an area of approximately 1 cm^2 and approximately 1 cm of stem cuttings with buds of tissue-cultured seedlings were used as explants.

3.2 Embryogenic callus induction

Embryonic calli were induced from young leaf lobe explants of tissue-cultured plantlets on somatic embryo induction medium (CIM) containing 12 mg L^{-1} picloram and were produced after 2–3 weeks. Embryogenic calli were picked out with tweezers and cultured continuously on CIM for 6–8 weeks for propagation. The medium was refreshed every 2 weeks. All cultures were kept in a growth chamber in the dark at 25°C. If the explants were stem cuttings with buds, they were cultured on axillary bud enlargement medium (CAM) for 2–4 days for bud enlargement before being cultured on CIM. The later cultures were the same as those used to culture young leaf lobe explants.

3.3 FEC induction

Embryogenic calli were cultured on GD for 2–4 weeks. The fine particles generated on the surfaces of embryogenic calli were separated and then propagated continuously on GD. FEC formed after 2–4 weeks of continuous circulation culture on GD. The medium was refreshed every 2 weeks. All cultures were kept in a growth chamber in the dark at 25°C.

3.4 Suspension culture

Cell suspension cultures were initiated by transferring approximately 1 g of FEC into a 100-mL flask with 30 mL of SH. The flask was agitated on a rotary shaker at 110–130 r min^{-1}. The liquid medium was refreshed every 2–3 days. All cultures were kept in a growth chamber at 25°C with a 12 h photoperiod and irradiance of 45 μmol^{-2} s^{-1}. Protoplast isolation was performed through 5 days of suspension culture in SH.

3.5 Protoplast isolation and purification

3.5.1 Enzymolysis

Large particles were removed from 5-day-old cell suspension cultures in SH with tweezers, and the liquid medium was aspirated out with a straw. Approximately 1 g of tissue was placed in a Petri dish (9 cm diameter) with 12 mL of cell digestion solution. The cell digestion solution contained a mixture of enzymes (10 g L^{-1} cellulase R-10, 400 mg L^{-1} macerozyme R-10, and 100 mg L^{-1} pectolyase from Yakult, Japan) and 1 mg L^{-1} NAA, 1 mg L^{-1} 2,4-D, 740 mg L^{-1} KNO_3, 368 mg L^{-1} $CaCl_2$, 34 mg L^{-1} KH_2PO_4, 492 mg L^{-1} $MgSO_4{\cdot}7H_2O$, 19.2 mg L^{-1} Na-EDTA, 14 mg L^{-1} $FeSO_4{\cdot}7H_2O$, 91 g L^{-1} D-mannitol, and 0.5 g L^{-1} MES. The suspension tissues were incubated in the enzyme solution for 18 h on a shaker at 40 r min^{-1} and 25°C in the dark.

3.5.2 Purification

Protoplasts were purified through the gradient centrifugation method. The digested tissues were filtered through a 45 μm stainless steel mesh to remove undigested cell clumps and debris. The filtrate was transferred into 10-mL centrifuge tubes and centrifuged for 6 min at 960 r min^{-1}. The supernatant was removed with a Pasteur pipet. The pellets were gently resuspended in 1.0–1.5 mL of 13% mannitol solution containing CPW nutrients (27.2 mg L^{-1} KH_2PO_4, 250 mg L^{-1} $MgSO_4$, 100 mg L^{-1} KNO_3, 150 mg L^{-1} $CaCl_2$, 0.2 mg L^{-1} KI, 0.003 mg L^{-1} $CuSO_4$). Then, the protoplast-containing 13% mannitol solution was slowly pipetted onto the top of 3–4 mL of 26% sucrose solution containing CPW nutrients while avoiding mixing and centrifuged for 6 min at 960 r min^{-1}. A band of viable protoplasts formed at the interface between the two layers. The protoplasts were carefully removed from the interface with a Pasteur pipet and resuspended in protoplast culture medium (TM2G). The protoplasts in TM2G were centrifuged for 6 min at 960 r min^{-1}. The supernatant was removed with a Pasteur pipet, and the protoplasts were resuspended in TM2G with 0.36 mol L^{-1} glucose at a density of 5×10^5 protoplasts mL^{-1}.

The yield of obtained protoplasts (cells g^{-1}) was calculated by using the following formula: $N \times 5 \times 10^4 \times V/m$; where N = number of protoplasts counted in a hemocytometer chamber; V = volume of diluted protoplasts; and m = fresh weight of plant material for protoplast isolation.

3.5.3 Viability test

The viability of the obtained protoplasts was checked with fluorescein diacetate (FDA). A total of 12 μL of 5 mg mL^{-1} FDA solution was added to 0.5 mL of the protoplast suspension. After 5 min, the protoplasts were examined with an Olympus IX71 inverted fluorescence microscope (green fluorescence, Olympus, Japan). The viability of obtained protoplasts (%) was calculated as follows: number of protoplasts with green fluorescence/Total protoplasts in the field × 100.

3.6 Protoplast culture

Initially, protoplasts were cultured through the thin liquid layer culture method in 1.5 mL of TM2G with 0.30–0.36 mol L^{-1} glucose in a 6-cm plastic Petri dish in the dark at 28°C. The medium was refreshed every 10 days: twice with TM2G with 0.30–0.33 mol L^{-1} glucose, then twice with a medium with reduced levels of glucose (0.27–0.30 mol L^{-1}). It was refreshed again two times with reduced glucose levels (0.25 mol L^{-1} glucose). The osmotic pressure of the culture was reduced by gradually reducing the glucose concentration of the TM2G medium to promote cell division.

3.7 Suspension culture

Protoplasts were cultured in TM2G with gradual dilution for approximately 6–10 weeks. Then, protoplast-derived compact calli were transferred into SH for suspension culture, and the other calli were cultured further. The liquid medium was refreshed every 7–15 days, and the calli were propagated in SH for 2–3 weeks.

3.8 Somatic embryogenesis

For embryo differentiation, the calli propagated in SH were transferred to MSN under light. The differentiated embryos were cultured on embryo maturation medium (CMM) for 1–3 weeks to develop large green cotyledon embryos. Then, the mature large green cotyledon embryos were transferred to shoot elongation medium (CEM). Shoot elongation began after 4–8 weeks. When the length of the elongated shoot reached 2–3 cm, the shoot was cut off for rooting on Murashige and Skoog (MS) medium. Rooting could occur in 7 days, and the protoplasts usually took 5–7 months to develop into complete plants.

3.9 Influencing factors of plant regeneration from cassava protoplast

3.9.1 State of FEC

The isolation of high-quality protoplasts is a prerequisite for protoplast culture, and the state of FEC directly affects the quality of isolated protoplasts, including yield and activity. FEC is characterized by a loose structure, the presence of spherical particles on its surface, and a milky white or yellow color. It can be used to establish suspension systems. When suspended in SH, numerous fine particles are dispersed.

In general, subculturing FEC on GD for 15–20 days results in FEC in the best state, i.e., friable and loose, and increases the yield to the maximum. After suspension culture, protoplasts are separated from FEC. This approach is conducive to plant regeneration. The protoplasts isolated from the embryogenic callus suspension of cassava subcultured for 5–15 days have high activity and few impurities.

3.9.2 Protoplast extraction and purification

The extraction and purification of cassava protoplasts separated in cell digestion solution are a key step. The cell digestion solution may not flow automatically when it is filtered through a stainless steel screen due to the effect of its surface tension, and filtration generally takes a long time. The longer the protoplasts stay in the enzyme solution, the lower their activity and the higher their impurity content. Therefore, the enzyme solution should be filtered through a stainless steel screen immediately. An external force can be exerted on the stainless steel screen to enable the enzyme solution to flow down and filter quickly. Purification through gradient centrifugation provides protoplasts with high yield and activity.

3.9.3 Protoplast culture

Cassava protoplasts were cultured in TM2G at a density of 5×10^5 pieces mL^{-1}. The initial concentration of glucose in TM2G can be within the range of 0.30–0.36 mol L^{-1}. Subsequently, the medium must be constantly refreshed and its glucose concentration must be reduced gradually to promote cell cluster division and growth. After protoplasts were cultured in TM2G for 6–8 weeks, 1–2 mm compact calli visible to the naked eye were selected, and other calli were used for further culture.

3.9.4 Embryo differentiation and germination

Sofiari et al. [35] believed that the differentiation of cassava protoplast-derived calli into embryos and the germination of embryos constitute the bottleneck of plant regeneration from protoplasts. Therefore, somatic embryogenesis is a key step in plant regeneration from cassava protoplast. The medium is an important factor in this process.

Given that the compact callus of protoplast origin is in the same state as the FEC used for cassava genetic transformation, compact callus of protoplast origin and FEC are considered as cell clusters, and the MSN used as the medium for genetic transformation is also used as the medium for embryo differentiation.

Before the differentiation of embryos on MSN, compact calli are first suspended in SH for 2–4 weeks. The compact calli become loose after being cultured in SH. This effect is advantageous for further somatic embryogenesis or proliferation on MSN or GD.

3.10 Composition and function of cassava culture medium

Nine kinds of cassava media are discussed in this chapter. **Table 1** shows the functions of nine kinds of media, and **Tables 1–3** present the composition of the nine kinds of media.

Cassava culture media containing MS, CAM, CIM, MSN, CMM, and CEM have basically the same compositions and differ only by hormone type or dosage. They are all composed of MS salt and vitamins, plus 20 g L^{-1} sucrose, 8 g L^{-1} agar, and 2 μM $CuSO_4$ (or not). They play different roles in the tissue culture and regeneration of cassava due to the different kinds or dosage of hormones that they contain (**Table 1**).

Although CIM and MSN media could be used to induce cassava embryos, they induce different explants. The explants often induced on CIM can be young leaf lobes, apical buds, and axillary buds used for the induction of primary, secondary, and circulating somatic embryos, which are in the beginning stages of somatic

Medium	Culture stage/function	Components
CAM	Axillary bud enlargement	MS salts and vitamins (**Table 2**), 2 μM $CuSO_4$, 10 mg L^{-1} 6-BA, 20 g L^{-1} sucrose, 8 g L^{-1} agar
CIM	Embryo induction	MS salts and vitamins (**Table 2**), 2 μM $CuSO_4$, 12 mg L^{-1} picloram, 20 g L^{-1} sucrose, 8 g L^{-1} agar
MSN	Embryo induction	MS salts and vitamins (**Table 2**), 2 μM $CuSO_4$, 1 mgL^{-1} NAA, 20 g L^{-1} sucrose, 8 g L^{-1} agar
CMM	Embryo maturation	MS salts and vitamins (**Table 2**), 2 μM $CuSO_4$, 0.1 mg L^{-1} 6-BA, 20 g L^{-1} sucrose, 8 g L^{-1} agar
CEM	Shoot elongation	MS salts and vitamins(**Table 2**), 2 μM $CuSO_4$, 20 g L^{-1} sucrose, 8 g L^{-1} agar, with 1.0 mg L^{-1} 6-BA
MS	Rooting and subculture of tissue culture seedlings	MS salts and vitamins (**Table 2**), 0.02 mg L^{-1} NAA, 20 g L^{-1} sucrose, 8 g L^{-1} agar
GD	FEC induction, maintenance, and proliferation	GD salts and vitamins, 12 mg L^{-1} picloram, 20 g L^{-1} sucrose, 8 g L^{-1} agar (**Table 3**)
SH	FEC suspension culture, maintenance, and proliferation	SH salts and vitamins, 12 mg L^{-1} picloram, 60 g L^{-1} sucrose (**Table 3**)
TM2G	Protoplast culture	TM-2 salts and vitamins, 1 mg L^{-1} NAA, 0.5 mg L^{-1} ZT, 54–64.8 g L^{-1} glucose (**Table 3**)

Table 1.
Medium components and function.

Macro-elements (g L^{-1})		Micro-elements (mg L^{-1})		Iron salts (mg L^{-1})		Vitamins (mg L^{-1})	
KNO_3	**19**	**KI**	**0.83**	**$FeSO_4{\cdot}7H_2O$**	**27.8**	**Glycine**	**2**
NH_4NO_3	16.5	H_3BO_3	6.2	$Na_2{\cdot}EDTA$	37.3	Myo-inositol	100
KH_2PO_4	1.7	$ZnSO_4{\cdot}7H_2O$	8.6	/	/	Nicotinic acid	0.5
$MgSO_4{\cdot}7H_2O$	13.7	$Na_2MoO_4{\cdot}2H_2O$	0.25	/	/	Pyridoxine HCl	0.5
$CaCl_2{\cdot}2H_2O$	4.4	$CuSO_4{\cdot}5H_2O$	0.025	/	/	Thiamine HCl	0.1
/	/	$CoCl_2{\cdot}6H_2O$	0.025	/	/	/	/
/	/	$MnSO_4{\cdot}H_2O$	16.9	/	/	/	/

Table 2.
MS salts and vitamins.

embryogenesis and can grow into different types of embryoids, such as globular, torpedo, and cotyledon embryos. The explants often induced on MSN are FEC or calli derived from protoplast division. When cultured on MSN under light, they can grow into different types of embryoids, such as globular, torpedo, and green cotyledon embryos. CIM is used as a somatic embryo induction medium under dark conditions, whereas MSN is used as a somatic embryo induction medium under light conditions.

The regeneration processes of globular, torpedo, and cotyledon embryos induced on CIM and MSN are the same. They all undergo and complete maturation, bud elongation, and rooting on CMM, CEM, and MS, respectively. These processes should be conducted under light conditions.

Medium components		GD	SH	TM2G
Macro-elements (mg L^{-1})	$Ca(NO_3)_2{\cdot}2H_2O$	208.81	/	/
	KCl	65.00	/	/
	KH_2PO4	300.00	/	170
	KNO_3	1000.00	2500	1500
	$MgSO_4$ $7H_2O$	35.00	400	370
	NH_4NO_3	1000.00	/	/
	$CaCl_2{\cdot}2H_2O$	/	200	440
	$NH_4H_2PO_4$	/	300	/
Micro-elements (mg L^{-1})	$CoCl_2{\cdot}6H_2O$	0.025	0.1	0.025
	$CuSO_4{\cdot}5H_2O$	0.025	0.2	0.025
	H_3BO_3	0.30	5.0	6.20
	KI	0.80	1.0	0.38
	$MnSO_4{\cdot}H_2O$	1.00	10	16.9
	$Na_2MoO_4{\cdot}2H_2O$	0.025	0.1	0.25
	$ZnSO_4{\cdot}7H_2O$	0.30	1.0	8.60
	$FeSO_4{\cdot}7H_2O$	0.278	15	13.9
	$Na_2{\cdot}EDTA$	0.336	20	18.5

Medium components		GD	SH	TM2G
Vitamins (mg L^{-1})	Glycine	4.00	/	0.50
	Myo-inositol	100.00	0.5	4600
	Nicotinic acid	1.00	0.5	2.5
	Pyridoxine HCl	1.00	0.1	1
	Thiamine HCl	10.00	/	10
	Folic acid	/	/	0.5
	Biotin	/	/	0.05
	D-Ca-pantothenate	/	/	0.50
	Choline chloride	/	/	0.10
	Casein hydrolysate	/	/	150
	L-cysteine	/	/	1
	Malic acid	/	/	10
	Ascorbic acid	/	/	0.50
	Adenine sulfate	/	/	40
	L-glutamine	/	/	100
	Riboflavin	/	/	0.25
Others (g L^{-1})	Sucrose	20	60	/
	Agar	8	/	/
	Glucose	/	/	54–64.8
	Mannitol	/	/	4.56
	Xylitol	/	/	3.80
	Sorbitol	/	/	4.56
	MES	/	/	0.098
Hormone (mg L^{-1})	Picloram	12	12	/
	NAA	/	/	1
	Zeatin	/	/	0.5

Table 3.
GD, SH, TM2G components.

GD can be used for the induction, maintenance, and proliferation of cassava FEC. SH can also participate in the maintenance of the embryogenic proliferation of FEC. GD is a solid medium, whereas SH is a liquid medium. FEC can be converted between GD and SH cultures, and its properties do not change. FEC cultured on SH has better cell consistency and faster proliferation than that cultured on GD. TM2G is used as a medium for culturing cassava callus protoplasts. Its osmotic pressure is reduced during culture by decreasing its glucose concentration gradually to promote cell division.

4. Conclusion

Cassava is a food crop, and the biotechnology research on cassava lags behind that on major food crops, such as rice and wheat. The establishment and development of cassava tissue culture and regeneration technology have promoted the application of biotechnology techniques, such as genetic transformation, genome editing, and somatic hybridization, to cassava. However, deficiencies remain. Cassava tissue culture and regeneration technology still need development and optimization to establish an efficient regeneration system without genotype dependence.

In cassava, protoplast regeneration technology could be applied to somatic hybridization and protoplast transformation. Somatic hybridization technology could break through the barriers of sexual hybridization and represents a direction for cassava breeding with protoplast regeneration technology as the basis. The disadvantages of cassava protoplast regeneration technology are high technical requirements and time consumption. We hope the chapter will be beneficial for the genetic improvement of cassava.

Acknowledgements

This work was supported by the Guangxi Scientific and Technological Development Subject, China (AB21196071), and the Basic Scientific Research Fund of Guangxi Academy of Agricultural Sciences, China (2021YT149).

Abbreviations

FEC	friable embryogenic callus
CAM	axillary bud enlargement medium
CIM	somatic embryo induction medium
MSN	somatic embryo emerging medium
GD	Gresshoff and Doy (1974) medium
SH	suspension culture medium
TM2G	protoplast culture medium
CMM	embryo maturation medium
CEM	shoot elongation medium
MS	Murashige and Skoog (1962) medium
2,4-D	2,4-dichlorophenoxyacetic acid
6-BA	6-benzylaminopurine
NAA	naphthaleneacetic acid

Author details

Wen Feng*, Hai-Tian Fu, Yan-Chun Luo and Jian-Qi Huang
Guangxi Subtropical Crops Research Institute, Nanning, P.R. China

*Address all correspondence to: wenfengw83@163.com

References

[1] Howeler R, Lutaladio N, Thomas G. Save and Grow: Cassava, a Guide to Sustainable Production Intensification. Rome: Food and Agriculture Organization of the United Nations; 2013

[2] Ojola OP, Nyaboga EN, Njiru PN, Orinda G. Overexpression of rice thaumatin-like protein (*Ostlp*) gene in transgenic cassava results in enhanced tolerance to *Colletotrichum gloeosporioides* f. sp. *manihotis*. Journal of Genetic Engineering & Biotechnology. 2018;**18**:125-131. DOI: 10.1016/j.jgeb.2017.12.002

[3] Ceballos H, Kulakow P, Hershey C. Cassava breeding: Current status, bottlenecks and the potential of biotechnology tools. Tropical Plant Biology. 2012;**5**:73-87. DOI: 10.1007/s12042-012-9094-9

[4] Parmar A, Sturm B, Hensel O. Crops that feed the world: Production and improvement of cassava for food, feed, and industrial uses. Food Security. 2017;**9**:907-927

[5] Mongomake K, Doungous O, Khatabi B, Fondong VN. Somatic embryogenesis and plant regeneration of cassava (*Manihot esculenta* Crantz) landraces from Cameroon. Springerplus. 2015;**4**:477. DOI: 10.1186/s40064-015-1272-4

[6] Jaramillo AM, Sierra S, Chavarriaga Aguirre P, Castillo DK, et al. Characterization of cassava ORANGE proteins and their capability to increase provitamin A carotenoids accumulation. PLoS One. 2022;**17**(1):e0262412. DOI: 10.1371/journal.pone.0262412

[7] Chavarriaga-Aguirre P, Brand A, Medina A, Prías M, Escobar R, Mar-tinez J, et al. The potential of using biotechnology to improve cassava: A review. In Vitro Cellular & Developmental Biology-Plant. 2016;**52**:461-478. DOI: 10.1007/s11627-016-9776-3

[8] Beyene G, Chauhan RD, Gehan J, Siritunga D, Taylor N. Cassava shrunken-2 homolog *MeAPL3* determines storage root starch and dry matter content and modulates storage root postharvest physiological deterioration. Plant Molecular Biology. 2022;**109**:1-17

[9] Bredeson JV, Lyons JB, Prochnik SE, Wu GA, Ha CM, Edsinger-Gon-zales E, et al. Sequencing wild and cultivated cassava and related species reveals extensive interspecific hybridization and genetic diversity. Nature Biotechnology. 2016;**34**:562-570. DOI: 10.1038/nbt.3535

[10] Brand A, Quimbaya M, Tohme J, Chavarriaga-Aguirre P. Arabidopsis *LEC1* and *LEC2* orthologous genes are key regulators of somatic embryogenesis in cassava. Frontiers in Plant Science. 2019;**10**:673. DOI: 10.3389/fpls.2019.00673

[11] Souza LS, Alves AAC, de Oliveira EJ. Phenological diversity of flowering and fruiting in cassava germplasm. Scientia Horticulturae. 2020;**265**:109253

[12] Liu J, Zheng Q, Ma Q, Gadidasu KK, Zhang P. Cassava genetic transformation and its application in breeding. Journal of Integrative Plant Biology. 2011;**535**:52-69

[13] Ceballos H, Kawuki RS, Gracen VE, Yencho GC, Hershey CH. Conventional breeding, marker-assisted selection, genomic selection and inbreeding in clonally propagated crops: A case

study for cassava. Theoretical Applied Genetics. 2015;**128**:1647-1667

[14] Kartha KK. Regeneration of cassava plants from apical meristems. Plant Science Letters. 1974;**2**:107-113

[15] Konan NK, Sangwan RS, Sangwan-Norreel BS. Efficient in vitro shoot-regeneration systems in cassava (*Manihot esculenta* Crantz). Plant Breeding. 1994;**113**:227-236

[16] Konan NK, Schöpke C, Cárcamo R, Beachy RN, Fauquet C. An efficient mass propagation system for cassava (*Manihot esculenta* Crantz) based on nodal explants and axillary bud-derived meristems. Plant Cell Reports. 1997;**16**:444-449

[17] Li HQ, Sautter C, Potrykus I, Puonti-Kaerlas J. Genetic transformation of cassava (*Manihot esculenta* Crantz). Nature Biotechnology. 1996;**14**:736-740

[18] Li HQ, Guo JY, Huang YW, Liang CY, Liu HX, Potrykus I, et al. Regeneration of cassava plants via shoot organogenesis. Plant Cell Reports. 1998;**17**:410-414

[19] Stamp JA, Henshaw GG. Somatic embryogenesis in cassava. Zeitschrift für Pflanzenphysiologie. 1982;**105**:183-187

[20] Raemakers CJJM, Amati M, Staritsky G, Jacobsen E, Visser RGF. Cyclic somatic embryogenesis and plant regeneration in cassava. Annals of Botany. 1993;**71**:289-294

[21] Raemakers CJJM, Schavemaker CM, Jacobsen E, Visser RGF. Improvement of cyclic somatic embryogenesis of cassava(*Manihot esculenta* Crantz). Plant Cell Peports. 1993;**12**:226-229

[22] Saelim L, Phansiri S, Netrphan S, Suksangpanomrung M, Narangajavana J. Optimization of in vitro cyclic somatic embryogenesis and regeneration of the Asian cultivars of cassava (*Manihot esculenta* Crantz) for genetic manipulation system. Global Journal of Biotechnology & Biochemistry. 2006;**1**(1):07-15

[23] Taylor NJ, Edwards M, Kiernan RC, Davey CDM, Blakesley D, Henshaw GG. Development of friable embryogenic callus and embryogenic suspension culture systems in cassava (*Manihot esculenta* Crantz). Biotechnology. 1996;**14**:26-730

[24] Syombua ED, Adero MO, Mbinda WM, Wanyonyi CN, Ngugi MP, Oduor RO. A system for rapid gene introgression into cassava immature leaves and subsequent recovery of transformed lines. Plant Biotechnology Reports. 2021;**15**:27-37. DOI: 10.1007/s11816-020-00654-1

[25] Lin HS, van der Toorn C, Raemakers KJJM, Visser RGF, De Jeu MJ, Jacobsen E. Development of a plant regeneration system based on friable embryogenic callus in the ornamental Alstroemeria. Plant Cell Reports. 2000;**19**(5):529-534. DOI: 10.1007/s002990050768

[26] Taylor NJ, Masona MV. Production of embryogenic tissues and regeneration of transgenic plants in cassava (*Manihot esculenta* Crantz). Euphytica. 2001;**120**:25-34

[27] Taylor N, Chavarriaga P, Raemakers K, Siritunga D, Zhang P. Development and application of transgenic technologies in cassava. Plant Molecular Biology. 2004;**56**:671-688

[28] Utsumi Y, Utsumi C, Tanaka M, Ha VT, Matsui A, Takahashi S, et al. Formation of friable embryogenic callus in cassava is enhanced under conditions of reduced nitrate, potassium and phosphate. PLoS One.

2017;**12**(8):e0180736. DOI: 10.1371/journal.pone.0180736

[29] Ibrahim AB, Heredia FF, Pinheiro CB, Aragao FJL, Campos FAP. Optimization of somatic embryogenesis and selection regimes for particle bombardment of friable embryogenic callus and somatic cotyledons of cassava(*Manihot esculenta* Crantz). International Journal of Histology and Cytology. 2020;**7**(16):2790-2797

[30] Elegba W, McCallum E, Gruissem W, Vanderschuren H. Efficient genetic transformation and regeneration of a farmer-preferred cassava cultivar from Ghana. Frontiers in Plant Science. 2021;**12**:668042. DOI: 10.3389/fpls.2021.668042

[31] Bull SE, Owiti JA, Niklaus M, Beeching JR, Gruissem W, Vanderschuren H. Agrobacterium-mediated transformation of friable embryogenic calli and regeneration of transgenic cassava. Nature Protocols. 2009;**4**:1845-1854. DOI: 10.1038/nprot.2009.208

[32] Wang YJ, Lu XH, Zhen XH, Yang H, Che YN, Hou JY, et al. A transformation and genome editing system for cassava cultivar SC8. Genes. 2022;**13**:1650. DOI: 10.3390/genes13091650

[33] Odipio J, Alicai T, Ingelbrecht I, Nusinow DA, Bart R, Taylor NJ. Efficient CRISPR/Cas9 genome editing of phytoene desaturase in cassava. Frontiers in Plant Science. 2017;**8**:1780

[34] Devang M, Alessandra S, Ravi BA, Shan-e-Ali ZS, Matthias H, Wilhelm G, et al. Linking CRISPR-Cas9 interference in cassava to the evolution of editing-resistant geminiviruses. Genome Biology. 2019;**20**:1-10

[35] Sofiari E, Raemakers CJJM, Bergervoet JEM, Jacobsen E, Visser RGF. Plant regeneration from protoplasts isolated from friable embroygenic callus of cassava. Plant Cell Reports. 1998;**18**:159-165

[36] Wen F, Xiao SX, Nie YM, Ma QX, Zhang P, Guo WW. Protoplasts culture isolated from friable embryogenic callus of cassava and plant regeneration. Scientia Agricultura Sinica. 2012;**45**:4050-4056. (In Chinese)

[37] Wen F, Su W, Zheng H, Yu B, Ma Z, Zhang P, et al. Plant regeneration via protoplast electrofusion in cassava. Journal of Integrative Agriculture. 2020;**19**(3):632-642

[38] Anthony P, Davey MR, Power JB, Lowe KC. An improved protocol for the culture of cassava leaf protoplasts. Plant Cell, Tissue and Organ Culture. 1995;**42**:229-302

[39] Mukami A, Juma BS, Mweu C, Ngugi M, Oduor R, Mbinda WM. Plant regeneration from leaf mesophyll derived protoplasts of cassava (*Manihot esculenta* Crantz). PLoS One. 2022;**17**(12):e0278717. DOI: 10.1371/journal. pone.0278717

[40] Shahin EA, Shepard JF. Cassava mesophyll protoplasts: Isolation, proliferation and shoot formation. Plant Science Letters. 1980;**17**:459-465

[41] Sofiari E. Regeneration and Transformation of Cassava. Netherlands: Wageningen Agricultural University of Wageningen; 1996

Chapter 5

Mutation Breeding: A Tool in Nutritional Improvement of Cassava

Amanze Ngozi Joan and Abah Simon Peter

Abstract

Cassava is an important food security crop worldwide with a lot of unexploited potential. More than 60% of global production is used for human consumption, while lesser quantity is used in livestock and Pharmacia industries. Improvement through hybridization and selection have been exploited but is limited by inter-specific and intra generic crop boundary, irregular flowering and low spontaneous mutation rate which cannot be depended on considering the high demand on the crop. Induce mutations has continue to remain an alternative tool for cassava improvement. The cytology analysis carried on five cassava varieties using varying levels of colchicine showed that the mutagen has significant aberration effect at ($p < 0.05$), with a Mitotic Index (MI) of (132.14), an error in cell divisions as shown in the positive increase yield of both parents and progeny of the cassava varieties evaluated. An epidermal-polyploidy change induced includes laggard, bridges, fragments, stickiness, vagrant and crises-cross at various concentrations. A required aberration was observed in the result. This shows significant difference in the mitotic index in a decreasing order with an increase in level of mutagen (132.14, 65.21 and 42.60) respectively. This result showed the mutagenic potentialities of colchicine in cassava induction and improvement.

Keywords: cassava breeding, induced mutation, karyotype, progenies, micronutrients, colchicine

1. Introduction

Cassava is a prominent root crop, which plays important role in the food security of many countries in the tropics, especially in sub-Sahara Africa. Every part of the plant is important but of most economic importance is the root. It is efficient in carbohydrate production and provides cheap source of calories for millions of people in Sub-Sahara. The largest producer of cassava in Africa is Nigeria and the third largest producer in the world after Brazil and Thailand [1]. Cassava is rated as the major staple crop in Nigeria feeding about 70% of the populace (FAOSTAT 2021). It has a fresh root starch content of about 30–40%, and the crop gives the highest yield of starch per unit area of both cereals and root and tuber crops [2, 3], but highly deficient in essential micronutrients, and extremely low in protein content which range between 1 and 3%. Although cassava is relatively rich in vitamin C, its content of iron, phosphorus, calcium and other minerals are in trace amount [4]. These micronutrient levels are not sufficient to take care of the micronutrient requirement of the low-income group of people who depend on root and tuber crops as their staple food. The inadequate intake of essential micro nutrients such as vitamin A, zinc and iron has been identified as the major causes of hidden hunger in the world practically in Nigeria. This has made malnutrition an immediate global challenge that requires urgent attention, and the global scale challenge posed by hidden hunger (micronutrient deficiency) informed its inclusion as one of the Sustainable Development Goals of the United Nations – SDG Goal 2: Zero Hunger, which has as one of its aims: to insure that enough nutritious foods are available to people by 2030 in a sustainable manner. Stephen in his health care nutrition analysis stated that malnutrition cannot be sustainably achieved through supplements and drugs, rather through a combination of options led by bio fortification of staple foods to produce whole and organic food; Food base approach is a more sustainable approach to attaining micronutrient adequacy compared to other methods [5, 6]. Bio fortification is a current complementary strategy to obtain and maintain adequate supply of essential micronutrients particularly among cassava products consumer.

2. The crop Cassava

i. *Morphological Description of Cassava*: (*Manihot esculenta*, with common names; cassava, manioc and tapioca Brazilian [7] is a procumbent and semi-herbaceous sub shrub plant in the family *Euphorbiaceae* (spurge family). It is native to South America but now grown in tropical and sub-tropical areas worldwide for the edible starchy roots. It can grow up to 7 m in height with a diameter of up to 20 cm, [8, 9], though most current varieties are hardly more than 3 m long. It is generally divided into two distant parts: the vegetative and the underground part. The vegetative part is made up of woody stem(s), which can be straight or branched. The outer bark is smooth, light brown to yellowish gray. The inner bark is cream-green with thin exudates, watery latex or sap but often bluish-gray when young. The branching pattern is typically dichotomous or trichotomous with branch-lets light green to tinged reddish having at the branching point a terminal inflorescence. Leaves are spirally arranged on petiole of up to 30 cm long, often reddish-purple. The root system is made up of the adventitious and the storage roots. The adventitious root is normally used for anchorage and absorption, while the storage root is the edible and the most important part of the plant. The tuberous edible root grows in clusters of 4–8 at the stem base. Roots are from 1 to 4 inches in diameter and 8–15 inches long, although roots up to 3 feet long roots up to 3 feet long have been found [10]. Some are usually cylindrical and tapered, some are irregular, while others are oval. They are covered with a thin reddish brown fibrous bark, which protects it from rodents and pest.

ii. *Karyotype and Origin of M. esculenta*: Cassava geographically is a native of America with centers of diversity in Brazil and Mexico [11]. The crop has more than 98 species which resulted from natural and artificial interspecific hybridization among M. esculenta and wide species [12]. Among these species, only *Manihot esculenta* Crantz is the most important and widely cultivated in the tropics and sub-tropical regions. This specie has a wide range of genetic diversity of more 7000 cultivars and 16,000 accessions due to the weak reproductive isolation barriers. This attribute makes transfer of desirable genes easy and manipulation of traits possible. Genetically cassava is a typical diploid having a chromosome number 2n = 36 with only few with more chromosome numbers [12].

iii. *Nutritional Composition of Cassava*: Cassava roots are very rich in carbohydrate. They contain significant amount of minerals - calcium (50 mg/100 g), phosphorus (40 mg/100 g) and vitamin C (25 mg/100 g) [1]. It is very rich in starch content, yielding more than 30% of starch per unit area [1]. However, they are poor in protein and other nutrients except the leaves, which are rich in protein (lysine), but deficient in other amino acids such as thiamine and tryptophan [13]. The protein content is extremely low and ranges between 1 and 3% [13, 14]. It has a large quantity of hydrocyanic, which makes it toxic, unhealthy to man, and animal. The quantity of this toxin depends on the variety, age of harvest and method of planting. However, this substance is very volatile and is removed to a safety level just by peeling, washing, grating and heating [14].

3. Production and utilization of Cassava in food and industrial

Global production and importance in the world economy: cassava is a universal crop produced either in subsistence level or in large scales. Production worldwide is contented in five countries namely Nigeria, Brazil, Thailand, Indonesia and Congo Democratic Republic cultivated under an average land area of 16.7 million hectares at the growth rate of 2.2%per annual Cassava. It is the most important staple food crop among the four major tropical root and tuber crops (cassava, yam, potatoes and cocoyam), providing basic diet for over half a billion people in the developing world [15, 16]. Globally, cassava is the second carbohydrate and starch source for food and industrial uses [17, 18] and the seventh most important food crop worldwide [19]. It is efficient in carbohydrate production and provides cheap source of calories for millions of people in Sahara, Sub-Sahara Africa. According to [20], more than 60% of global production is used as food for man, with lesser quantity being used for animal feed and agro based industries. In sub-Saharan Africa and Latin America, cassava is mostly used for human consumption [21], whereas in Asia and parts of Latin America, it is mostly used commercially for the production of animal feed and starch-based products [22]. Nigeria is the largest producer of cassava in the world producing more than

i. The roots can be processed into various forms of starch for domestic consumption, local and foreign market. It can also be utilized fresh, as in the case of sweet cultivars (low cyanorganic glycosides) or in processed forms as flour, starch and animal feed in the case of bitter cultivars (high cyanogenic glycosides [23]. Cassava fresh leaves are rich in crude protein (21.39%) and are utilized for human food as vegetable or as a constituent in the form of source eaten alone or with main staple [24]. The stems are used for propagation, staking and as firewood. In animal nutrition, it is either used in feed formulation or eaten fresh. The cassava water known as Manipulearia in Brazil is used for animal fattening and enhancement of milk production in dairy farming (unpublished lecture). They are also used in the production of local gin and liquor. Different countries make different types of alcoholic beverages from cassava: Caum and tiquira Brazil, nihamanchi South America, impata Mozambique and others. Cassava-based dishes are widely consumed wherever the crop is cultivated, and the food forms are either regional, or national [25]. Cassava has been found as source of alternative energy that is strong, renewable and sustainable. In some economies such as China, it has gradually become a major source of ethanol production (Business Green news, 2008). In addition to this, China-based Hainan Yedao Group invested $51.5 m (£31.8 m) to produce 33 million US gallons (120,000 m^3) a year of bio- ethanol from cassava plants in 2008 [26]. It is an attractive fuel crop because it can give high yields of starch and total dry matter in spite of drought conditions and poor soil.

ii. Cassava production and utilization in Nigeria: Cassava (Manihot esculents crantz) cassava is staple crop of choice across Nigerian households, playing very significant role in the diets and income of the producing households [27] with less production cost per unit output than any other staple food crop, it is capable of growing well and giving reasonable yield on marginal soils because of its drought-tolerance Cassava is grown in virtually all the states in Nigeria but more prominently grown in the following states Benue, Kogi, Enugu, Imo, Cross-River, Ondo, Ogun, Delta, Anambra, Edo, and Taraba, States [28, 29]. It is the congregate of production from these areas that placed Nigeria the largest producer of cassava worldwide producing an estimate of 52million per annual Cassava is the subject of many expansion programmes in the Sub-Saharan African region, as commercializing cassava and do-mystically producing staple crops – in order to limit imports. It– remains a key objective of many West African governments. In Nigeria, the regional production leader, the "Anchor Borrower's Programme" (ABP), initiated by the country's Central Bank (CBN), and provides preferential loans to smallholder farmers who provide their product to the processing sector. However, while cassava is one of the many commodities listed in the programme, the implementation of ABP has, in effect, made rice more lucrative to cultivate. More so, the annual increase in demand for cassava flour in Nigeria to feed her local and foreign companies for bread production, sugar based additive, glues, plywood, textiles, paper, monosodium glutamate, drugs, bio-degradable products and as bio- energy production has made cassava production an economic necessity [28].

4. Cassava: a potential raw material in animal feed industries

Cassava is an ideal alternative crop as a source of energy in livestock industry. It has been used in various forms to feed livestock worldwide. All parts of the crop plants-the leaves, the stem and roots can be fed to animals processed or wholly, singly or mixed with other stuffs depending on the type of animal. For ruminates the whole plants can be chopped sun dried and fed. For the monogastric they are processed into palates or feed meals. Cassava can replace about 30 to 50 percent of corn in animal feed ration. However, this crop capable of providing high potential source of energy in animal feed production is limited by a number of problems, which reduces its use and utilization. The most common are the poor quality feed outcome of cassava mills due to lack of essentials micronutrients in cassava raw materials which calls for the use of additives (vitamins-minerals and amino acids.) additional cost, high moisture content, the present of toxic substance hydrocyanic acid and order. Presently several researches on the use of cassava as energy source replacing maize has shown that the above mention problems can be corrected through development and improvement programs which encourage the development of bio fortified cassava varieties that will supply those essential nutrients in a whole meal [29].

4.1 Cassava improvement

Early cassava breeding programs have largely focused on increasing cassava productivity and resistance to pest and disease through inter-specific hybridization, which led to the release of many varieties [30]. Later in the 70s focused on enhanced food qualities and traits for utility, which led to the release of high dry starch and dry matter content [31]. However it was identified that conventional breeding present's limitation in the breeding of specific traits needed to move cassava utilization forward [32]. This therefore led to the development of other breeding technologies which could manipulate on the genes and genomes to identify gene of great importance to farmers such as Obasanjo, gain changers, and others in Nigeria m. Some advanced breeding platforms have also developed breeding pipelines for enhanced cassava nutrition – IITA and NRCRI breeding for enhanced pro-vitamin A. Allard [33] also identified wild species with high protein content and with the advancement in plant breeding techniques, the problems of poor micronutrients and other issues such as adaptation to arid and semi-arid conditions are being addressed [33–36].

Some conventional and unconventional breeding approaches have been implode in the improvement – Several conventional breeding approaches implode in the improvement of cassava are hybridization which used inter-specific crosses in cassava breeding to develop recombinants between cultivated and wild cultivars from which clones with better characters were obtained Storey and Nichols. Through this breeding approach many hybrid cassava cultivars that are resists to prevailing disease and pest, have be developed and released in many countries [36]. However, hybridization and selection have been exploited but cannot work beyond the biological boundary of inter-specific and intra generic crop breeding, its time consuming, their high degree of long vegetative propagation with its associated low and irregular flowering, high heterozygosity and difficulty of selection of recombinant and the problem of transferring undesirable traits and un locking desirable traits in sterile crops could not always be dependable due to gene actions and involved for the trait and diverse genetic structure of the parent in case of brake down. These necessitate the exploration of other breeding tools in order to improve difficult traits in crops with high degree of long vegetative propagation or sterility.

4.2 Mutation breeding

Mutation breeding despite the few bottle necks associated with it was the earliest tool used by plant breeders to increase plant size, develop non exiting traits and generate variations in crops [37]. Plants are made up of genes, which are the molecular unit of heredity of every living organism. The nucleus of an organism's cell contains a number of having normally (2n sets) of chromosomes of which if they appear in pairs, the organism is said to be in a diploid, while and if more than a pair, the organism any organism that has higher sets of chromosomes is regarded as a polyphoid [11].

Polyploidy, an accidental change in the cytology of an organism, can be brought about by spontaneous or induced mutation. Spontaneous Mutation breeding is a breeding tool that can be used to create genetic variation for traits that generally have

Figure 1.
Cassava root generated from the treated materials.

low variability in seed and vegetative propagated crops. It has a great potential in genetic improvement of cassava, cocoyam and cereal such as rice and ray and other root and tuber crops. But Spontaneous Mutation breeding, however begin low cannot meet the ecological, industrial and economical challenge of time led to the diversification of technologies to increase cassava productivity (**Figure 1**).

(iii) Induction Mutation: In recent times, induction mutation and analyses of mutants have received great attention as alternative means of crop improvement [12]. Breeding for micro-nutrients in cassava- In order to produce cassava varieties with high micronutrient levels, different breeding methods and tools have been used such as biotechnology (genetic engineering and molecular breeding techniques), and tissue culture approaches such as soma clonal variation and somatic hybridization. Induction mutation using colchicine is an excellent improvement tool successfully used in the manipulation of plant genome for the development of traits in sterile and irregular flowering plants of importance, including in roots and tuber crops [38] (**Figure 2**).

(a) (b)

Figure 2.
(a) Typical Variety TMS0505 plant (b) Manipulated/induced variety TMS0505 plant.

5. Creation of genetic diversity using mutagne Colchicine

However, there is still immense scope to enhance the mineral nutrient, essential amino acid altered protein and fatty acid profiles, physicochemical properties of starch, phyto-nutrients, reduce anti-nutritional factors in cassava for human and animal consumption through bio fortification [39]. Induced mutation was conducted in National Root Crops Research Institute (NRCRI), Umudike Research farm, with Five cassava varieties namely: TMS98/0505, TMS94/4479, TMS98/1632, TMS92/0057, TMS98/0581 treated with manipulating hormone at three levels of colchicine: 0, 2 and 4 ppm. (Parts per million) (**Figure 3**).

The stakes were soaked in the solution for 30 minutes, air dried for 24 hours in the screen house, pre-sprouted in nursery bags and transplanted to the field at 3-leaf stage. At 7 months after planting, 25 pieces of 25 cm stake cuttings each of the cassava varieties were cut from the mature plants raised from colchicine treated materials and planted in a well harrowed and ridged field in a 4×5 randomized block design, at a spacing of 1×1 m intra and inter-row replicated three times.

The cytology evaluation was screened and calculated for chromosomal aberration using the following formulae:

$$\text{Mitotic Index}(MI) = \frac{\text{Number of dividing cell}}{\text{Total number of cells counted}} \quad (1)$$

$$\%\text{Abberrant cells} = \frac{\text{Total aberrant cell}}{\text{Dividing cells}} \times 100 \quad (2)$$

$$\text{Mitotic inhibition} = \frac{\text{Mitotic index of control} - \text{Mitotic index of treated}}{\text{Mitotic index of control}} \times 100 \quad (3)$$

Figure 3.
(a) Cassava plant with capsule or fruit (b) Seeds harvested from induced parents.

Conc (ppm)	No. of dividing cells	Chromosome aberration per 1000 cells						Total aberrant cells	Mitotic index	% aberrant cells	Mitotic inhibition
		Laggard	Bridge	Fragment	Stickiness	Vagrant	C-mitosis				
0 ppm	660	0	0	0	0	0	0	0	132.14 ± 8.83	0.00 ± 0.00	0.00 ± 0.00
2 ppm	325	5	11	6	17	3	3	45	65.21 ± 8.18*	0.19 ± 0.01	0.51 ± 1.83*
4 ppm	213	7	19	7	24	5	7	69	42.60 ± 9.54**	0.32 ± 0.03	0.68 ± 1.71*

** and ** mean significant at 5% and 1% respectively.*

Table 1.
Cytological effects of colchicine on cells of Manihot esculenta.

The proximate analyses of the roots were carried out using AOAC (1990) methods. Observations made showed that the treatment with colchicine mutagen had significant effect on the sizes of the stomata with and physico composition of the parent and their progenies. The microscopic analysis showed that the mutagen has significant aberration effect on the varieties across the concentrations ($p < 0.05$), with a Mitotic Index (MI) value of (132.14) and this led to error in cell divisions as shown in the positive increase yield of both parents and progeny of the cassava varieties evaluated, while there was no chromosomal aberration in the control. The type of change induced by the colchicine in this study was epidermal-polyploidy change which includes laggard, bridges, fragments, stickiness, vagrant and crises-cross at various concentrations. Mutation frequency calculated reported significant difference in the mitotic index in a decreasing order with the increase in level of mutagen (132.14, 65.21 and 42.60) respectively as shown in **Table 1**.

This result was advantageous in the induction of required changes in the studied cassava varieties and showed the mutagenic potentialities of colchicine. There was no significant difference among the three levels of concentration in most the physico-chemical compositions evaluated, but the concentration level 4 ppm gave highest ash content followed by level 2 and 0 ppm (2.437, 2.50 and 2.63%) respectively. On the other hand crude fiber was significantly affected by concentrations of colchicine as seen in the result (2.25, 2.46 and 2.65) and increased with increased level of colchicine from 0 to 4 ppm level. The starch content of the progenies evaluated according to levels (0, 2 and 4 ppm) were significantly different at ($p < 0.05$) with level 4 ppm higher than other levels (32.00, 29.44 and 34.03%) respectively, with an average of 31.70%, a value comparable to those of [36], who reported an average starch content of 32.6. Concentration level increased the major minerals of interest: zinc, iron and Magnesium significantly. Both zinc and iron were significantly affected by concentration level 4 ppm, while. Magncsium content at concentration level 2 ppm (0.58 mg/100 g) significantly differed from the other two concentration levels at ($p < 0.05$) of 4 ppm (0.48 mg/100 g) than followed by concentration level 0 ppm or control (0.41 mg/100) This result simply indicates that these essential nutrients can be enhanced using induced mutation and that the concentration levels has not been reached. The other mineral composition of the cassava materials evaluated: Nitrogen,

Figure 4.
Cassava root generated from F1 seeds.

Calcium, Potassium, Sodium and Phosphorous were not significantly affected. Some vitamins and amino-acids were significantly different among the three levels of concentrations evaluated. Concentration level 4 ppm affected hydrogen cyanide higher than other levels, followed by concentration level 2 ppm. While concentration level 2 ppm affected Phenol more than other levels (0.26 mg/100), followed by concentration level 4 ppm (0.17 mg/100), than 0 ppm (0.13 mg/100). Although Vitamin C was higher than other vitamins across level (22.51, 22.52, 22.51 mg/100 g), it was not significant. Thiamin, nicotinic, riboflavin (25.51, 22.51, 22.52 mg/100 g) was not only significant, there by proving that they were genetic in nature and can be improved through conventional breeding methods (**Figure 4**).

6. Conclusion

Since the progressive increase level of colchicine continued to increase the level of some micro nutrients but did not in others, it is concluded that the nutritional values of cassava can be improved through mutation breeding using colchicine and suggest that level of application has not been exploited. Secondly, the fact that the present increase were lower than the Recommended Dietary Allowance of these nutrients [4], where the mean value range for zinc (1.08–2.52 mg/100 g) is low, iron (9.95–13.87 mg/100 g) is high, potassium (716.91–757.68 mg/100 g) is low, sodium (42.46–80.85 mg/100 g) magnesium (89.68–128.35 mg/100 g) is average, calcium (22.77–30.73 mg/100) is low and phosphorus (42.60–45.89 mg/100 g) there is still need to continue to increase the concentration of colchicine used in cassava improvement while animal sources be used as complements. For animal feed production, with enhanced botanical seed production through induced mutation, continuous and selection of variables for values of enhanced feed quality such as low moisture, low peel fiber in addition with the fortified micronutrient quality the goal of substituting higher percentage maize for animal feed production will be actualized.

Author details

Amanze Ngozi Joan[1*] and Abah Simon Peter[2]

1 Department of Crop Production, National Root Crops Research Institute, Umudike, Abia State, Nigeria

2 Department of Biotechnology, National Root Crops Research Institute, Umudike, Abia State, Nigeria

*Address all correspondence to: amanzengozi@gmail.com

References

[1] Oriola KO, Raji AO. Trends at mechanizing cassava postharvest processing operations. International Journal of Engineering and Technology. 2013;**3**(9):879-889

[2] Tonukari NJ. Cassava and the future of starch. Electronic Journal of Biotechnology. 2004;**7**:1-4. DOI:10.2225/vol7-issue1-fulltext-9

[3] Salcedo A, Del Valle A, Sanchez B, Ocasio V, Ortiz A, Marquez P, et al. Comparative evaluation of physiological post-harvest root deterioration of 25 cassava (*Manihot esculenta*) accessions: Visual vs. hydroxycoumarins fluorescent accumulation analysis. African Journal of Agricultural Research. 2010;**5**:3138-3144

[4] Enidiok SE, Attah LE, Otuechere CA. Evaluation of moisture, total cyanide and fiber contents garri produced from Cassava (Manihot utilissima) varieties obtained from Awassa in Southern Ethiopia. Pakistan Journal of Nutrition. 2008;**7**(5):625-629

[5] Low JW, Arimond M, Osman N, et al. A food-based approach introducing orange-fleshed sweet potatoes increased vitamin A intake and serum retinol concentrations in young children in rural Mozambique. African Journal of Food, Agriculture, Nutrition and Development. 2007;**22**:21397-21417. DOI: 10.18697/ajfand.113.19835

[6] USDA GRIN Taxonomy. Retrieved 4 January 2014

[7] Healthline Media UK Ltd, Brighton, UK, 2004-2021 a Red Ventures Company. All rights reserved. MNT is the registered trademark of Healthline Media

[8] Nassar NMA, Elkholy H, Eltantawy A. Thirty years of word cassava production. In: Intern. Symp. on Tropical Root and Tuber Crops, Souvenir program. Thiruvananthapuram: Central Tuber Crops Research Institute; 2000. pp. 5-13

[9] Nassar NMA. Cassava *Manihot esculenta* Cratz genetic resources: Their collection, evaluation and manipulation. Economic Botany. 1999;**69**:179-230

[10] Dixon AGO, Asiedu R, Bokanga M. Breeding of cassava for low cyanogenic potential: Problems, progress and perspectives. Acta Horticulture. 1994;**375**:153-162

[11] Song C, Liu S, Xiao J, He W, Zhou Y, Qin Q, et al. Polyploid Organisms. Science China Life Sciences. 2012;**55**:301-311. DOI: 10.1007/s11427-012-4310

[12] Buitrago AJA. La yuca en la alimentation animal. Cali, Colombia: Centro International de Agricultural Tropical (CIAT); 1990. p. 446

[13] Heuze V, Tran G, Bastianelli D, Archimede H, Lebas F, Regnier C. Cassva tubers.feedipedia.org a programme by INRA, CIRAD, AFZ and FAO. 2012B. Available from: http//www.feedipededia.org/node/527

[14] FAO (Food and Agriculture Organization of the United Nations). Food energy-methods of analysis. 1995

[15] Ravindran V. FAO Animal Production and Health Paper. Vol. 95. Rome, Italy: Food and Agriculture Organization of the United Nations; 1992. pp. 111-125

[16] Shu QY. A summary of the international symposium on induced mutations in plants. In: Shu QY, editor. Induced Plant Mutations in the Genomics Era. UN, Rome: FAO; 2009. pp. 15-18

[17] Huang J, Bachem C, Jacobsen E, Visser RGF. Molecular analysis of differently exposed genes during post-harvest deterioration cassava (*Manihot escluenta*) by sexual seeds. Euphytica. 2001;**120**:85-93

[18] Thakkar SH, Huo T, Maziya-Dixon B, Faille ML. Impact of style of processing on retention and bio-accessibility of Carotene in Cassava (Manihotesculentacrantz). Journal of Agric Food Chem. 2009;**57**:1344-1348

[19] Nwokoro SO, Orheruata AM, Ordiah PI. Replacement of maize with cassava sievates in cockerel starter diets: Effect on performance and carcass characteristics. Tropical Animal Health and Production. 2002;**34**(2):163-167

[20] Nyerhovwo JT. Cassava and the future of starch. Electronic Journal of Biotechnology. 2004;**7**:1-8

[21] FAO, Rome FAOSTAT. Food and Agricultural Commodities Production. 2011. Available from: http://faostat.fao.org [accessed 31 January, 2013]

[22] Balagopalan C. Cassava utilization in food, feed and industry. In: Hillocks RJ, Tresh JM, Bellotti AC, editors. Cassava Biology, Production and Utilization. New York, USA: CABI Publishing; 2002. pp. 209-23521

[23] Frederick DO. Hog and Hominy Soul Food from Africa to America. Chapters 1-2. Columbia University Press; 2008. pp. 1-16. Available from: http://www.jstor.org/stable/10.7312/opie14638.5 [Accessed: December 22, 2023]

[24] Sayre R, Beeching JR, Cahoon EB, Egesi C, Fauquet C, Fellman J, et al. The BioCassava plus program: Biofortification of Cassava for Sub-Saharan Africa. Annual Review of Plant Biology. 2011;**62**:251-272

[25] Wossen T, Girma G, Abdoulaye T, Rabbi I, Olanrewaju A, Bentley J, et al. The Cassava Monitoring Survey in Nigeria Final Report. Ibadan, Nigeria: IITA; 2017. 66 p. ISBN 978-978-8444-81-7

[26] FGN. Cassava Master planA Strategic aAction Plan for Development of Nigerian Cassava Industries. 2006

[27] Otim-Nape GW. Epidemiology of the African cassava mosaic Gemini virus disease (ACMD) in Uganda. [Phd thesis] University of Reading, UK. 1993. p. 256

[28] Ceballos H, Iglesias AC, Perez JC, Dixon AGO. Cassava breeding: Opportunities and challenges. Plant Molecular Biology. 2004;**56**:506-516

[29] Akimbo AO, Okogbenin E, Adebija C, Ewa F, Ekaette E, Ikeorgu U, et al. Development of high yielding early cassava varieties for drought prone agro-ecology of Nigeria: International annual meetings of the American society of Aronomy, crop science of America, and soil sciences. Fundamental for Life: Soil, Crop, & Environmental Science ASA. CSSA. SSSA. International Annual meetings October 16 – 19, 2011, San Antonio, TX; 2011

[30] Nassar NMA, Dorea G. Protein contents of cassava cultivars and it's hybrid with *Manihot* species. Turrialba. 1981;**32**(4):429-432

[31] Amanze NJ, Eke-Okoro ON, Ikeorgo JG, Ohaneje T. Influence of variety, micro–nutrient and growth hormone on some reproductive and yield parameters of Yam. In: Proceedings of the Agricultural Society of Nigeria: October 16th–20th 2006. 2006. pp. 374-379

[32] Kawano K. Thirty years of cassava breeding for productivity: Biological and social factors for success. 2003

[33] Allard RW. Principles of Plant Breeding. New York: John Wiley and Sons; 1960. 485 p

[34] Cromarty AS, Ellis RH, Oberts EH. The Design of Seed Storage Facilities for Genetic Conservation IBPGR, Rome Italy. Bioversity International; 1982. p. 96

[35] Adepoju OT, Nwangwu JO. Nutrient composition and contribution of noodles (abacha) and local salad from cassava (Manihot spp) to nutrient intake of Nigerian consumers. African Journal of Food Science. 2010;**4**(7):422-426

[36] Strauss SH, Costanza A, Séguin A. Genetically engineered trees: Paralysis from good intentions. Science. 2015;**349**:794-795. DOI: 10.1126/science.ab0493

[37] Mba C, Afza R, Jankowicz-Cieslak J, Bado S, Matijevic M, Huynh O, et al. Enhancing genetic diversity through induced mutagenesis in vegetatively propagated plants. In: Shu QY, editor. Induced Plant Mutations in the Genomics Era. Rome, Italy: Food and Agriculture Organization of the United Nations; 2009. pp. 293-296

[38] Cebballos H, Sánchez T, Denyer K, Tofiño AP, Rosero EA, Dufour D, et al. Induction and identification of a small-granule, high-amylose in cassava (*Manihot esculenta* Crantz). Journal of Agricultural and Food Chemistry. 2008;**56**(16):7215-7222. DOI: 10.1021/jf800603. Epub 2008 Jul 26. PMID: 18656938

[39] Chavarriaga-Aguirre P, Brand A, Medina A, Prías M, Escobar R, Martinez J, et al. The potential of using biotechnology to improve cassava: A review. In Vitro Cellular & Developmental Biology - Plant. 2016;**52**(5):461-478. DOI: 10.1007/s11627-016-9776-3

Chapter 6

Improving Cassava Cultivation as an Industrial Raw Material on Acid Soil in Indonesia

Bariot Hafif, Yulia Pujiharti, Alvi Yani, Noveria Sjafrina, Robet Asnawi, Nendyo Adhi Wibowo, Andri Frediansyah, Neneng Laela Nurida and Ai Dariah

Abstract

Cassava is grown nowadays for use in food, feed, and industrial purposes. It is believed that the agro-industrial sector, which uses cassava as a raw material, has more advanced farming technology for improving cassava production. Lampung province in Sumatra Island, Indonesia, is one of the cassava production centers for industrial raw materials, with a planted area of 256,632 ha in 2018. The planting areas are acid soils of Ultisols, Inceptisols, and Oxisols with pH levels ranging from 4.5 to 5.0. Acidic soils have a complicated set of plant growth-limiting constraints. Essential nutrients for plant growth, such as N, P, and K, as well as other cations, are often low due to leaching, nutrients fixed by Fe/Al oxides of clay minerals, and low soil cation exchange capacity. In these acid soils, cassava production ranges from 8 to 15 t ha^{-1} for traditional farming, 20–24 t ha^{-1} for semi-developed farming to 25–35 t ha^{-1} for advanced farming. Meanwhile, with numerous technological advancements, cassava productivity can reach 40–50 t ha^{-1}. Aside from improving varieties, technological updates being pursued include increasing the accuracy of mineral fertilizer dosage, improving planting system technology, bio-fertilizer technology, and *in situ* organic C enrichment of acid soils.

Keywords: acid soil, cassava varieties, fertilizer, the plantation system, biofertilizer, Indonesia

1. Introduction

Cassava (*Manihot esculenta* Crantz) is the third staple food after rice and corn in Indonesia, having carbohydrate content 40% higher than rice and 25% higher than corn. Cassava has a starch content of about 24%, which gives it potential as a raw material for bioethanol [1]. It can be grown on marginal land and has a high tolerance to acid soil [2]. These advantages make cassava farming possible with no or low input to the soil.

Cassava cultivation exists throughout Indonesia, such as in large islands like Sumatra, Java, Kalimantan, Sulawesi, and Papua, and small islands like Bali,

Sumbawa, and Maluku, etc. Cassava is suitable to be developed in the wet tropical climate that dominates the territory of Indonesia and also grows quite well in the dryer part of Indonesia, like on the island of Nusa Tenggara [3]. Under high temperatures, high light intensity, and heavy rainfall in Indonesia, cassava for industrial raw materials requires a long maintenance time until harvest, about 9 to 10 months [4].

In Indonesia, planting cassava as a food ingredient is rarely expanded by farmers in large areas. Farmers grow cassava as a secondary crop, usually on narrow land (several hundred square meters), and in some areas, the cassava yield sometimes goes into food stocks as the mixing of rice during famine periods like at the end of the dry season or the beginning of the growing season. In places where cassava is grown as an industrial raw material, it is more common to grow it over a large area (> 1 ha). Lampung Province, in the southern part of Sumatra Island, Indonesia, is one of the production center areas. In Lampung, most farmers sell cassava yields to factories that process them for tapioca flour, feed, and bioethanol. At present, there are around 130 units of cassava processing factories in this area, with a demand of about 5 million tons of cassava per year [5]. In 2018, Lampung's total cassava farming land was around 295,548 ha. Other areas of Indonesia that develop cassava farming extensively are East Java Province, with approximately 157,899 ha, and Central Java Province, with 155,660 ha [6].

The type of soil for cassava farming in the three provinces of cassava production centers is quite different. In the provinces of Central Java and East Java, farmers are growing cassava on Alfisols, and Inceptisols, with slightly acidic to neutral soil pH (pH 5.0–6.5) [4, 7, 8], Meanwhile, in Lampung province, the soil is dominated by Ultisols, Oxisols, and Inceptisols, with soil pH ranging from 4.5 to 5.0 [9, 10].

2. Acid soil and constraints for plant growth in Lampung

Mulyani et al. [11] reported that acidic soils develop very widely in Indonesia, especially in wet climates, such as on the island of Sumatra. One of the areas on the island of Sumatra with extensive acidic dryland is the province of Lampung. The acidic dryland in this area reaches approximately 2.87 million ha and is dominated by the orders of Oxisols, Ultisols, and Inceptisols. Oxisols consist of the great group of Hapludox and Kandiudox, Ultisols consist of the great-groups Hapludult and Kanhapludult, while Inceptisols consists of the great-groups Dystrandept, Dystropept, and Eutropept [12]. The classification of acid soils in the Lampung area is following the soil classification as the soil classification of the USDA Taxonomy [13]. The profile of the prime acid soil orders found in the Lampung and their general properties are presented in **Tables 1–3**.

The low availability of phosphorus in tropical acid soils is due to P chelation by clay minerals, namely by amorphous and crystalline hydrous oxides of Fe and Al clay minerals, which is very conducive to happening in low pH soil. As shown in **Figure 1**, the forms of P chelated by clay minerals in acid soils are $H_2PO_4^-$ and HPO_4^{2-} [20]. The availability of K is low in acid soils, especially in those that have undergone advanced weathering, such as Ultisols and Oxisols. Rainfall and high temperatures speed up the release of K from rocks and other parent materials into the soil solution. Then, heavy rain keeps washing K out of the soil as low exchangeable K range of 0.03–0.11 meq 100 g^{-1} in the acid soils of Lampung in **Table 4**. In the study of cassava

Soil profile	Horizon	Depth (cm)	Additional information
	Ap; topsoil	0–7	Oxisols are one of the acidic soils that spread widely in the province of Lampung [14]. It is mature soil, formed by desilication and laterization processes in humid and warm tropical climates, and is heavily weathered. In the lower layer or horizon B, Oxisols have an oxic horizon (Bo) containing large amounts of hydrous-oxide or sesquioxide clay and kaolinite clay. Other Oxisols have plinthite, a deep red-yellow-gray or white mottled horizon. Because of the electrical charge, the Oxisols are sometimes variable charge soils [15]. The soil with an average pH of 4.5 or less, low nutrients, low available P, and low K^+, Ca^{2+}, Mg^{2+}, and high H^+, Al^{3+} is a great group of Kandiudox found in the central area of Lampung [16].
	AB; transition horizon, with organic matter	7–30	
	Bo; oxic horizon with a high concentration of Fe and Al oxides and hydroxides	30–104	
	BC; transition horizon to parent materials	>104	

Table 1.
The Oxisol profile and properties in Lampung Province, Sumatra Island, Indonesia [9].

Soil profile	Horizon	Depth (cm)	Additional information
	Ap; topsoil	0–14	Ultisols acid soils in the Lampung area are mature soils formed by a combination of laterization and podsolization processes, with an argillic horizon (a relatively impermeable clay-rich horizon) and accumulated concretion (c) in the sub-horizon (Bt). The great group of Ultisols included Kanhapludult, Kandidult, and Plinthudult, with a pH range of 4.4–4.8 [9]. High P fixation and low exchangeable K content were the main limiting factors for plant growth in this soil [16, 17]. Another great group of Ultisol, Kandiudult, was found in Central Lampung. This soil has high clay and low nutrient content, and soil acidity is in the pH range of 5.0–5.4 [16].
	AB; sub-soil with organic matter	14–37	
	Btc; an argillic horizon as an accumulation of leached clay (t), and concretion (c)	37–110	
	BC; transition horizon to C	>110	

Table 2.
The Ultisol profile and properties in Lampung Province, Sumatra Island, Indonesia [9].

Soil profile	Horizon	Depth (cm)	Additional information
	Ap; topsoil	0–24	The acid soil of Inceptisols in Lampung is soil with moderate weathering [18], with the main characteristic being the presence of a cambic horizon, a soil horizon whose structure developed weakly but indicated chemical transformation and material displacement (leaching). With a pH range of 4.5–5.5, inceptisols typically react from acidic to slightly acidic. In this soil, nutrient reserves such as P and K and other exchangeable cations vary from low to high, and base saturation tends to be high [19].
	Bw; cambic horizon, with weakly structured development	24–80	
	C; parent materials	>80	

Table 3.
The Inceptisol profile and properties in Lampung Province, Sumatra Island, Indonesia (private document of Hafif).

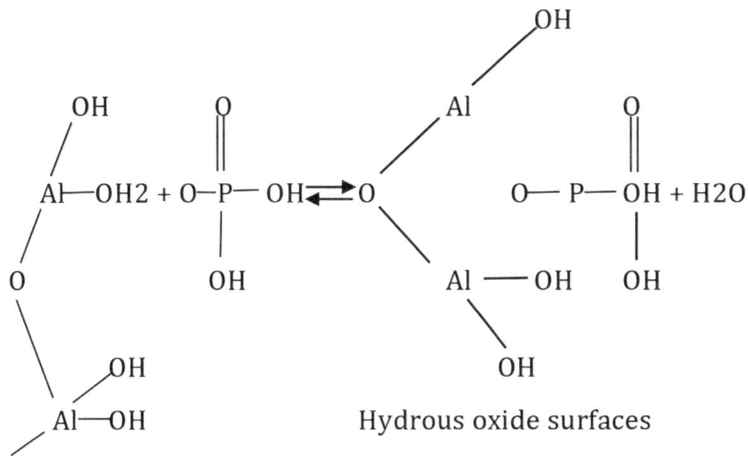

Figure 1.
Phosphorus (P) fixation in acidic tropical soil by amorphous and crystalline hydrous oxides of aluminum (Al) clay minerals. Source: Basak and Rakshit [20].

cultivation on acid soil, the field experiments were conducted on four great groups of two soil orders, Ultisol and Oxisol. The great group of Oxisol was Hapludox, and the three great groups of Ultisol were Plinthudult, Kandiudult, and Kanhapludult. The great group of Oxisol, Hapludox, is the Oxisol has soil moisture regime udic (soil in a humid climate), with no other identifying horizons. The great groups Plinthudult are sub-order Udult, with plinthite (mixture of clay with other minerals riching iron and humus-poor) found in 150 cm horizon from the soil surface. The Kandiudult is a sub-order Udult of the Ultisols, having a kandic horizon, and the Kanhapudult is the other sub-order Udult having a kandic horizon [13]. The general physicochemical characteristics of the acid soils used for cassava study in the field are presented in **Table 4**.

3. Plantation systems, fertilizer, and cassava production

Generally, there are two systems in cassava plantations in Lampung Province, namely monoculture and intercropping. The study by Manihuruk et al. [21] found the factors influencing the farmers in choosing a plantation system were land area, the distance of the farming area from the processing factory, and the source of income. Farmers having small areas (< 1 ha) prefer the intercropping system because it generates more revenue than the monoculture system. Farmers with farming areas near processing plants prefer the monoculture system because the cost of production transport to the factory is relatively low. Meanwhile, farmers with other sources of income choose an intercropping system because the earnings from the monoculture cassava are lower than those from intercropping, especially if they have small areas (≤ 0.5 ha).

The planting distance of cassava in a monoculture system widely used by farmers is 1 m x 1 m or 1 m x 0.8 m. While in the intercropping system, the distance of cassava between double rows was 160 cm, and in double rows, it was 80 cm. In an intercropping system, the first step is to grow seasonal crops such as soybeans, peanuts, or corn on land. Then the cassava planting is carried out after the seasonal crops are 15–30 days old.

Soil characteristics	Hapludox	Plinthudult	Kandiudult	Kanhapludult
Textures				
• Sand (%)	39	50	11	17
• Silt (%)	13	14	17	48
• Clay (%)	48	36	72	35
pH				
• H_2O	4.4	4.6	4.4	4.8
• KCl	3.9	4.2	4.1	4.3
Organic C (%)	1.28	1.16	1.51	1.40
Total N (%)	0.09	0.09	0.11	0.11
C/N	14	13	14	13
P_2O_5 (ppm)	9.3	5.3	23.2	13.0
K_2O (HCl 25%) (mg 100 g^{-1})	2	4	2	6
Exchangeable Cations (1 N NH_4OAc at pH 7)				
• Ca meq 100 g^{-1}	1.30	1.75	3.00	2.84
• Mg meq 100 g^{-1}	0.39	1.02	0.81	0.51
• K meq 100 g^{-1}	0.03	0.08	0.03	0.11
• Na meq 100 g^{-1}	0.05	0.09	0.07	0.05
CEC meq 100 g^{-1}	6.04	5.02	8.56	6.65
BS (%)	29	59	46	45
Al^{3+} meq 100 g^{-1}	1.26	0.47	0.71	0.13
H^+ meq 100 g^{-1}	0.26	0.21	0.17	0.13
Al saturation (%)	71.1	16.0	18.2	4.3

Source: Hafif [9], CEC=Cations exchange capacity, BS=Base saturation.

Table 4.
The physicochemical characteristics of some great-group acid soils at a soil depth of 0–20 cm in Lampung.

Based on cassava farming patterns such as planting system, inputs to the land, seedling kind, and land area, Hafif [9] reported that there were three types of cassava farming in Lampung; traditional, semi-developed, and advanced. Farmers in the traditional type typically work on 1 ha or less of cassava land, using a monoculture or intercropping system, cultivating the land with family labor and livestock, using random seeds (derived from previously planted cassava), and using very little mineral fertilizer. The input to the land is only manure at a modest rate of around 1–1.5 t ha^{-1}, and sometimes a little urea is added ($\pm$100 kg ha^{-1}). Another habit is often harvesting too-young cassava due to pressing economic needs. When converted to hectares areas, traditional farmer's land produced cassava of around 8–15 t ha^{-1} (**Table 5**). Most of them consider cassava products as additional/side income.

Farmers in the semi-developed farming category prefer a monoculture system and have used complete chemical and organic fertilizers. They mostly used compound fertilizers (NPK 15:15:15%) with doses varying between 50 and 200 kg ha^{-1} and supported by manure ranging from 1 to 4 t ha^{-1} in a cassava planting area of 0.5–2 ha.

Variables	Farmers of		
	Traditional	**Semi-developed**	**Advanced**
Cultivated area	< 1 ha	0.5–2 ha	2–5 ha
Cassava seedling	random seedlings	random seedlings- superior varieties	superior varieties
Organic fertilizer	1–1.5 t ha^{-1}	1–4 t ha^{-1}	2.5–4 t ha^{-1}
Chemical fertilizers	Urea 100 kg ha^{-1}	Compound fertilizer, NPK (15:15:15): 50–200 kg ha^{-1}	Compound fertilizer, NPK (15:15:15): 300–500 kg ha^{-1}
Soil Cultivation	family labor and or own livestock	Livestock or tractor services	Full tractor services
Cassava Production	8–15 t ha^{-1}	20–24 t ha^{-1}	25–35 t ha^{-1}

Source: Hafif [9].

Table 5.
Comparison of traditional, semi-developed, and advanced cassava farming characteristics in Lampung Province, Indonesia.

For soil cultivation, the farmers used livestock or tractor services (hand tractor) and still relied on random seeds. Cassava production from semi-developed farming for monoculture patterns ranges from 20 to 24 t ha^{-1} (**Table 5**).

In advanced cassava farming, the farmers have considered the efficiency and effectiveness of a farm. In cultivating the land, farmers have fully used the services of a tractor to cultivate the land in a shorter amount of time, about 1–2 hours per ha or 7 ha per day, whereas it would take five days per ha to do the same work with livestock. Most cassava seeds are derived from outside the land, based on recommendations from extension workers or factories. The area of cassava planting ranges from 2 to 5 ha. Advanced farmers preferred the monoculture system and used fully compound fertilizers (NPK 15:15:15%), sometimes adding urea and SP-36. The amount of compound fertilizer applied is high, between 300 and 500 kg ha^{-1}, and manure as much as 2.5–4 t ha^{-1}. Advanced farmers harvest cassava until it is 9–10 months old, with cassava production ranging from 25 to 35 t ha^{-1} (**Table 5**).

4. Cassava varieties as industrial raw matter

Cassava of UJ-5, UJ-3, Malang 4, Malang 6, Ardira 2, Ardira 4, and Litbang UK 2 were among the new high-yielding varieties introduced by the Indonesian Ministry of Agriculture in Lampung since 2000 [4]. One of the varieties that are becoming a favorite and being developed by many farmers in Lampung is UJ-5. The Indonesian Legumes and Tuber Crops Research Institute was the inventor of the superior cassava varieties, especially for industrial raw materials.

The UJ-5 variety could meet the requirements as a fuel-grade ethanol (FGE) raw material, due to having the following properties; 1) high starch content, 2) high yield potential, 3) resistance to biotic and abiotic stresses, and 4) flexibility in farming and harvesting time [22]. According to the Indonesian Legumes and Tuber Research Institute, the UJ-5 could produce tubers in the range of 25–38 t ha^{-1}, had starch content of 20–30% fresh weight (FW), and HCN content >100 ppm (slightly bitter taste) and harvest age of 9–10 months. UJ-5 was relatively resistant to cassava

Cassava clones	Water content	Dry weight	Starch	Sugar	Amy lose	Tubers for 1 L bioethanol	Yield
	(%)	(%)	(%)	(%)	(%)	Kg	t ha^{-1}
UJ-3	58.8 d	40.8 b	25.7 c	36.7 b	21.3 bc	4.9	25
UJ-5	54.8 e	43.9 a	27.5 a	39.1 a	22.4 a	4.5	30
Adira 4	64.5 a	37.7 c	22.6 f	31.0 d	18.8 e	4.7	17
Malang 6	60.3 c	40.4 b	23.5 e	34.1 c	20.8 c	5.0	15.6
MLG3011	59.7 c	40.7 b	24.3 d	33.3 c	21.0 c	4.3	27
CMM99023–4	64.5 a	37.6 c	21.9 g	30.3 d	20.1 d	5.1	25.3
CMM99008–3	58.4 d	40.2 b	26.9 b	37.0 b	21.7 b	4.2	22.3
OMM9908–4	61.7 b	40.1 b	24.5 d	34.1 c	19.8 d	4.5	31.7
LSD	0.8	1.1	0.46	0.91	0.53		

Column means followed by the unequal letter are significantly different at an LSD of 0.05. Source: Ginting et al. [22].

Table 6.
Comparison of some new high-yielding varieties and the properties of cassava as industrial raw materials and bioethanol in Lampung, Indonesia.

bacterial blight (CBB) and had high dry matter (% DM), starch content from dry matter (% DM), sugar content (% FW), and amylose (%) of which was 43.9, 27.5, 39.1, and 22.4, respectively, and the conversion of fresh tubers to bioethanol was 4.5 kg liter^{-1} (**Table 4**) [22, 23]. Other beneficial properties of the UJ-5 are; 1) leaves do not fall quickly, 2) it can grow on low and high pH soils, 3) it can grow in high populations, and 4) they can develop in an intercropping system [24]. **Table 6** shows in more detail the benefits of the UJ-5 compared to several other types of industrial raw materials and bioethanol.

5. Cassava yield quality

The low available nutrient content in acid soils, especially P and K, was the cause of low cassava quality due to low starch content and high cyanogenic glucosides [25]. Another cause of low starch content was harvesting cassava before maturity (at 6–7 months old), which was common among farmers in traditional and semi-advanced farming. On average, the starch content of the cassava grown on the acidic soil of Lampung is around 18–22% (manufacturer's information), although the potential starch content of the UJ-5 variety can reach 30%. However, cassava yield factories may accept these starch levels and limit cassava purchases to only those with a starch content is at least 18% [9].

The low quality of cassava causes the price of this commodity to fluctuate. From 2011 to 2016, the average cassava price decreased by around 2.38% per year [6]. The decline in prices caused some cassava farmers to switch farming to other commodities resulting in a reduction in the cassava planting area of 10.8% per year in Lampung [5]. However, with improvements in cultivation technology and growing superior varieties, cassava productivity was indicated to increase. In 2018, the cassava productivity of Lampung Province was about 26.04 t ha^{-1}, which was better than the average national productivity of 24.39 t ha^{-1}, and the selling price of cassava at the farmer level has continued to improve [6].

6. Technology improvement of cassava cultivation on acid soil

6.1 Mineral fertilizers

The study by Wargiono [26] on the acid soil in Lampung found that cassava in an intercropping system with rainfed rice gave the best result with the application of 90 kg N, 50 kg P_2O_5, and 90 kg K_2O per ha. The results of Ernawati's research [27] on Kanhapludult acid soil found the application of a mixture of urea, SP36, and KCl fertilizer in a ratio of 2:1:1 or the equivalent of a mix of 90 kg N: 36 kg P_2O_5: 60 kg K_2O with an application dose starting at 40 g $plant^{-1}$, 80 g $plant^{-1}$, 120 g $plant^{-1}$, and 160 g $plant^{-1}$ or the equivalent of 400 kg ha^{-1}, 800 kg ha^{-1}, 1200 kg ha^{-1} and 1600 kg ha^{-1} if the cassava population was 10,000 ha^{-1}, the yield of cassava was not significantly different, namely 54 kg $plant^{-1}$ or 54 t ha^{-1}. That means the lowest dose of mixed fertilizers, 400 kg ha^{-1}, was sufficient for cassava planted on a ha of acid soil.

KCl application of as much as 300 kg ha^{-1} on acid soil in Lampung increased the weight of cassava tubers by an average of 1.98 kg $plant^{-1}$ compared to an average of 1.45 kg $plant^{-1}$ by application of 200 kg KCl ha^{-1} [28]. Meanwhile, the study of Hafif [9] found that the application of straw compost (2 t ha^{-1}), each enriched with 50 kg KCl, 100 kg KCl and 200 kg KCl, to acid soil in Lampung, significantly increased the weight of tubers of cassava from 7.35 kg $plant^{-1}$ (without enrichment) to 7.97, 8.26 and 8.42 kg $plant^{-1}$ (**Table 4**), respectively, and the same treatments increased tuber starch content from 30.1% to 30.9%, 32.3%, and 33% (FW), and decreased total cyanogen content by 13.8%, 26.4%, and 28% from 246 ppm (**Table 7**).

To increase cassava production on acid soils, it is necessary to solve some problems such as Al toxicity [29], low P and K availability [2], and aggregate instability due to low soil organic matter content [30]. Although cassava is a tolerant plant for marginal lands, without fertilizer application, the yield of cassava was far from the target. Even soil fertility under cassava plants will rapidly decline due to the high nutrient uptake

The treatments	Stem diameter (cm)	Tuber weight (kg $plant^{-1}$)	Tuber number $plant^{-1}$	Starch content (%FW)	Total cyanogen (ppm)
K0	2.39	7.35 b	22.7	30.1 d	246 a
K50	2.52	7.97 a	25.8	30.9 c	212 b
K100	2.55	8.26 a	25.1	32.3 b	181 c
K200	2.57	8.42 a	25.2	33.0 a	177 c
LSD		0.97		2.11	26.9

The column means followed by the unequal letter are significantly different at an LSD of 0.05. Source: Hafif [9].

Table 7.
The application effect of straw compost (2 t ha^{-1}) enriched by 50, 100 dan 200 kg KCl on stem diameter, tuber weight, tuber number, starch, and the total cyanogen of cassava in acid soil in Lampung.

of cassava [31]. Howeler [32] reported the plantation of cassava for eight years consecutively without fertilization, which caused cassava production to decrease from 22 tons ha^{-1} to 13 tons ha^{-1}. Therefore, to get a high yield of cassava on marginal land, one should apply sufficient NPK and organic matter [25, 30]. Among the macronutrients, the K mineral is the one that plays a principal role in increasing the quantity and quality of cassava in acid soils [4, 24].

6.2 Improvement of the plantation system

Intercropping and monoculture systems are two options for cropping systems developed and used by Lampung cassava farmers. Research by Asnawi and Arief [33] found that cassava productivity could increase if the monoculture cropping system was changed to a monoculture with a double-row system. The monoculture system with a double row was different from the monoculture system of farmers, especially in terms of spacing, population per hectare, and fertilization rate. In the monoculture system of farmers, the spacing varies, namely 60 x 70 cm, 70 x 80 cm, or 80 x 80 cm. The total population of cassava in this monoculture system ranges from 15,000 to 20,000 plants per ha^{-1}. Fertilization is usually only 75–100 kg of urea plus a little SP-36 (50 kg ha^{-1}) and manure of about 1 t ha^{-1}.

The monoculture system with double rows changed the spacing to 160 x 80 x 80 cm so that the cassava plant population per hectare is only around 11,200 plants. This system recommended a fertilizer dose of 100 kg Urea +150 kg NPK + 100 kg KCl and manure to be 5 t ha^{-1}. Cassava yields can reach 50–60 t ha^{-1} with this system [33]. In addition, the system with double rows can join the intercropping system by planting annual crops in the 160 cm space between the double rows. This method is even considered more profitable. The first step in the intercropping system with double rows was to grow seasonal crops such as soybeans, corn, peanuts, and green beans on the space between double rows (160 cm), then plant cassava when the crop was two weeks to a month old. The performance of the intercropping system with double rows and corn as an intercropping crop is shown in **Figure 2**.

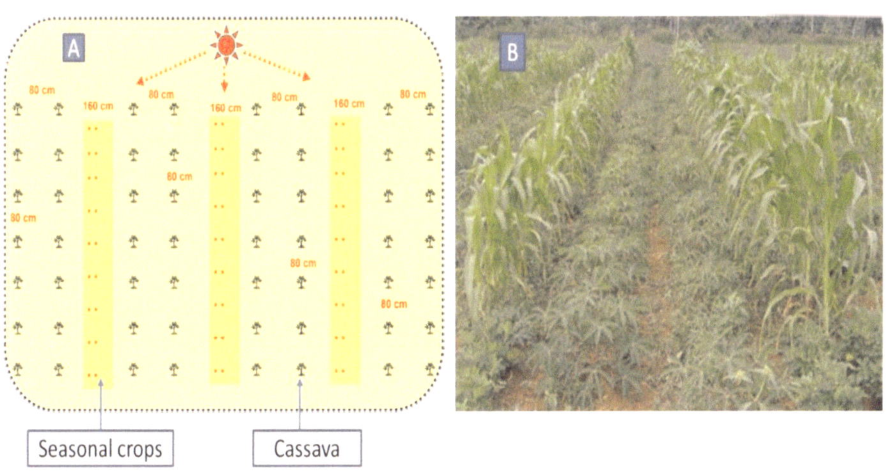

Figure 2.
Design of the intercropping double-rows system (a) and performance of the intercropping double-rows system in the field (B). Source: Robet and Arief [33].

6.3 Bio-fertilizer of mycorrhizae (arbuscular mycorrhiza)

According to Howeler [34], another ingredient that also had the potential to improve the growth, yield, and yield quality of cassava is *Arbuscular mycorrhizae* (AM). Cassava can grow well on acid soils with low P contents because it has a very efficient symbiosis with AM, which occurs naturally. Cassava is most dependent on AM. At low concentrations of P in acidic soils, the growth and branching of AM hyphae will increase. The AM performed a symbiosis with the cassava roots, as illustrated in **Figure 3** [9].

One way of enriching soil mycorrhiza is through the application of biofertilizers. When AM and plant roots form a symbiotic mutualism, the plant roots will supply exudate to AM, and vice versa, AM will help deliver nutrients and water to the roots. AM hyphae will extend the root system of plants up to 100 times and help plants absorb more nutrients and water, especially in soil with less available nutrients like P and microelements like Zn, Mo, and Cu. AM also increases plant tolerance to drought, high temperatures, infections from fungal diseases, and even high soil acidity. Good plant growth with the help of AM is easier to see in the crops planted in acid soils with a high level of weathering, low base cations and P, and high Al content [35].

According to the findings of Hafif [9], the use of AM bio-fertilizer combined with zeolite as a carrier was able to enhance cassava yield from 7.1 kg plant^{-1} to 8.8 kg plant^{-1}, and the amount of starch produced increased from 29.5% to 32.1% (FW) or 75.1% to 76.7% (DW) (**Table 8**).

6.4 In situ enrichment of soil organic C with root exudates of Brachiaria

The research on degraded soils in Madagascar showed that Brachiaria grass, as the source of nutritious feed for livestock in the tropic, planted as an intercrop between cassava, had a good effect on cassava production, namely being able to increase cassava yield from 4 to 13 t ha^{-1} to 11–30 t ha^{-1} or an average increase of 240% [36]. The beneficial effects of Brachiaria root exudates include their ability to improve soil aggregates, nutrient cycles, and organic carbon levels [36–38].

A study conducted by Hafif [9] demonstrated that the roots of signal grass (*Brachiaria decumbens*) released low molecular weight organic acids into the

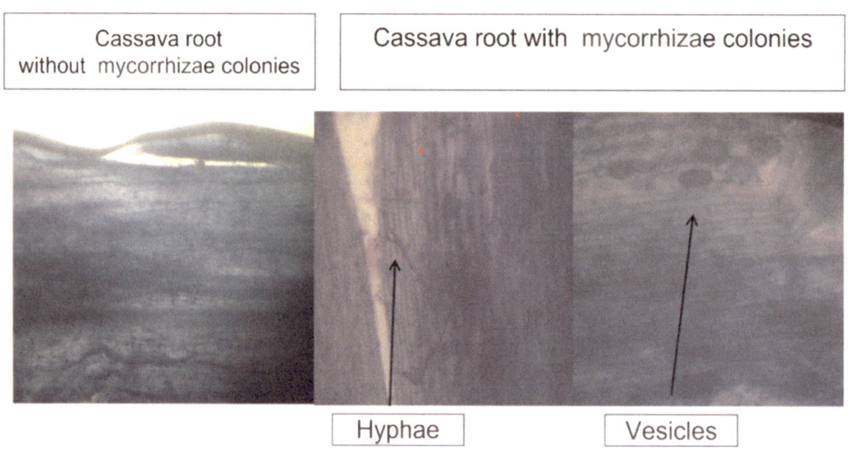

Figure 3.
Symbiosis mutualism between arbuscular mycorrhiza *(AM) and cassava roots. Source: Hafif [9].*

Treatment	Tuber weight (kg plant^{-1})	Tuber number plant^{-1}	Starch content (% FW)	Total cyanogen (ppm)
M0	7.1 b	22.9	29.4 b	210
M1	8.8 a	25.5	32.1 a	198
LSD	0.85		2.13	

The column means followed by the unequal letter are significantly different at an LSD of 0.05. Notes: FW = fresh weight, M0 = no mycorrhiza, M1 = with mycorrhiza.
Source: Hafif [9].

Table 8.
The effect of AM bio-fertilizer on quantity and quality of cassava yield on acid soils.

rhizosphere. These acids included citric, malic, and oxalic. When compared to results obtained without the presence of root exudate from Brachiaria grass, the organic acids secreted by Brachiaria roots were able to chelate aluminum with a significantly higher organic aluminum content. Research on acid soil in Lampung found the root exudates of Brachiaria could reduce the amount of exchangeable aluminum by up to 33%. A decrease in exchangeable Al by root exudates will increase P mobilization in acid soil by inhibiting P fixation by the Al oxide-hydroxide adsorption complex [39]. Planting Brachiaria grass as an intercrop between cassava on acid soils in Lampung (**Figure 4**)

Figure 4.
Brachiaria grass performance as an intercrop between cassava in field lab (A) and farmer's field (B) on acid soil in Lampung, Indonesia. Source: Hafif [9].

Treatments	Tuber weight (kg plant^{-1})	Tuber number plant^{-1}	Starch content (%) (FW)	Total cyanogen (ppm)
BD0	7.1 b	22.9	29.2	214 a
BD1	8.2 a	24.7	31.0	195 b
LSD	0.85			19.0

The column means followed by the unequal letter are significantly different at an LSD of 0.05. Source: Hafif [9].

Table 9.
The effect of Brachiaria root exudates on the yield quantity and quality of acid soil Lampung Indonesia.

increased yield and cassava starch. Brachiaria grass increased cassava tuber weight from 7.1 kg plant^{-1} to 8.2 kg plant^{-1} and starch content from 29.2% to 31.0% (FW) and reduced total cyanogen from 214 ppm to 195 ppm (**Table 9**).

7. Conclusion

Cassava in a tropical climate like Indonesia is one of the principal food sources, especially for marginalized people in rural areas. However, in certain areas, such as Lampung Province, cassava, which initially received little attention from farmers, has instead developed into one of the leading commodities. This positive development started in 2005, along with the rapid development of cassava processing factories in this region [5].

Farmers in Lampung are not discouraged from growing cassava because of the acidic soil. Cassava planting on acid soil with little external input could produce around 8–15 t ha^{-1}. However, if the next planting still has low input, then cassava production will decrease because cassava absorbs soil nutrients very highly. Based on that experience and supported by intensive counseling from the factory officer and agricultural extension from the local and central governments, the way farmers cultivate cassava is improving. In semi-developed and advanced cassava farming, cassava can produce 20 to 35 t ha^{-1}.

In Lampung, the average productivity of cassava is still around 17.53 t ha^{-1} [5]. This productivity is far from optimal because the experimental results can reach 40–50 t ha^{-1}. The productivity of cassava on acid soils can increase if farmers improve or adopt cultivation technologies such as planting superior varieties, increasing the doses of mineral fertilizers and organic fertilizers, and improving cropping systems. In the future, it is necessary to encourage the use of biological fertilizers of mycorrhiza, organic C enrichment, and increased mobilization of soil nutrients in situ by planting intercrops that produce root exudates like Brachiaria among cassava plants.

On the other hand, the slow absorption of cassava cultivation technology in Lampung was due to several factors. One of the most influential is the unstable and relatively low selling price of cassava at the farmer level. Low prices make it difficult for farmers to survive in cassava farming. As a result, from 2011 to 2016, the cassava planting areas in Lampung decreased by 10.8% per year because farmers switched their farming to other commodities. However, since 2018, the price of cassava has continued to improve, and this is the hope that farmers will get excited again about growing cassava in Lampung [6].

Author details

Bariot Hafif[1]*, Yulia Pujiharti[2], Alvi Yani[3], Noveria Sjafrina[3], Robet Asnawi[4], Nendyo Adhi Wibowo[5], Andri Frediansyah[5], Neneng Laela Nurida[6] and Ai Dariah[1]

1 Research Center for Horticulture and Plantation, National Research and Innovation Agency (BRIN), Cibinong Bogor, Indonesia

2 Research Center for Macroeconomics, and Finance, National Research and Innovation Agency (BRIN), Jakarta, Indonesia

3 Research Center for Agroindustry, National Research and Innovation Agency (BRIN), Indonesia

4 Research Center for Behavioral and Circular Economic, National Research and Innovation Agency (BRIN), Jakarta, Indonesia

5 Research Center for Food Technology and Processing (PRTPP), National Research and Innovation Agency (BRIN), DI. Yogyakarta, Indonesia

6 Research Center for Food Crops, National Research and Innovation Agency (BRIN), Cibinong Bogor, Indonesia

*Address all correspondence to: bari002@brin.go.id

References

[1] Tonukari NJ. Cassava and the future of starch. Biotechnology issues for developing countries. Electronic Journal of Biotechnology. 2014;7(1):5-8. [Online]. Available: http://www.academicjournals.org/AJB

[2] Howeler RH. Cassava cultivation and soil productivity. In: Clair H, editor. Achieving Sustainable Cultivation of Cassava. Burleigh Dodds Science Publishing Limited; 2017. pp. 258-300

[3] Nugraha HD, Suryanto A. Study of the potential productivity of cassava (Manihot esculenta Crant.) in Pati. Produksi Tanam. 2015;**3**(8):673-682

[4] Saleh N, Taufiq A, Widodo Y, Titiek S. Pedoman Budi Daya Ubi Kayu Di Indonesia. In: Taufiq A, Nasir S, Gusyana D, editors. 1st ed. IAARD Press; 2016. ISBN 9786023441358

[5] Mukhlis J, Hendra R, Asnawi RW, Arief A, Irawati, Purnamayani R. "Double Row, Bikin Kaya Petani Singkong Lampung". In: From Zero To Hero. 1st ed. Munier FF, Handoko S, Nurwullan E, Sirnawati E, Eds. Bogor: Balai Besar Pengkajian dan Pengembangan Teknologi Pertanian (BBP2TP); 2021. pp. 113-118

[6] BPPT. Outlook Teknologi Pangan 2018; In: Adiarso, Ismariny, Jaizuludin, Prihawantoro S, Supriyanto, editors. Pusat Pengkajian Industri Proses dan Energi (PPIPE). Jakarta. 2018. ISBN 9786021328064

[7] Wijanarko A, Sudaryono, Sutarno. Karakteristik Sifat Kimia dan Fisika Tanah Alfisol di Jawa Timur dan Jawa Tengah. Iptek Tanaman Pangan. 2007; **2**(2):214-226

[8] Kumalasari SW, Syamsiyah J. Studi beberapa sifat fisika dan kimia tanah pada berbagai komposisi tegakan tanaman di sub das solo hulu. Ilmu Tanah dan Agroklimatologi. 2011;**8**(2): 119-124

[9] Hafif B. Peningkatan kualitas tanah masam dan hasil ubikayu dengan brachiaria, mikoriza dan kompos jerami padi diperkaya kalium. Bogor: IPB; 2011

[10] Hafif Ernawati B, Pujiarti Y. Opportunities for increasing the productivity of smallholders oil palm. Jurnal Littri. 2014;**20**(2):100-108

[11] Mulyani A, Sarwani M. Karakteristik dan Potensi Lahan Sub Optimal untuk Pengembangan Pertanian di Indonesia. Sumberdaya Lahan. 2013;**7**(1):47-55

[12] Sukarman R, Hafif B, Hidayat H. Peta Zona Agroekologi Provinsi Lampung Skala 1:250.000. Jakarta: Bandar Lampung; 2013

[13] Soil Survey Staff. Keys to Soil Taxonomy. 12th ed. Vol. 12. Washington DC: USDA-NRCS; 2014

[14] Mulyani A, Rachman A, Dariah A. Penyebaran Lahan Masam, Potensi Dan Ketersediaannya Untuk Pengembangan Pertanian. In: Sastramihardja H, Manalu F, Aprillani SE, editors. Fosfat Alam: Pemanfaatan Fosfat Alam yang Digunakan Langsung Sebagai Pupuk Sumber P. Balai Penelitian Tanah Bogor. 2009. pp. 25-46. ISBN 9786028039192

[15] Tan KH. Book reviews: Environmental soil science. Geoderma. 2001;**99**:329-331

[16] Sutriadi MT. Pemupukan K Tanaman Padi Gogo pada Tanah Oksisol Kandik di Lampung Tengah. Tanah dan Iklim. 2017;**41**(2):91-100

[17] Nurida NL, Dariah A. Increasing upland Rice productivity in acid soil through due to the application of biochar soil Ammendment. Penelitian Pertanian Tanaman Pangan. 2019;**3**(2):67-74. DOI: 10.21082/jpptp.v3n2

[18] Prasetyo BH. Tanah merah dari berbagai bahan induk di indonesia: prospek dan strategi pengelolaannya. Sumberdaya Lahan. 2009;**3**(1):47-60

[19] Nursyamsi D, Budiarto A, Anggria L. Pengelolaan Kahat Hara pada Inceptisols untuk Meningkatkan Pertumbuhan Tanaman Jagung. Tanah dan Iklim. 2002;**20**(1):56-68

[20] Basak BB, Rakshit R. Chemistry of macronutrients fixation in acidic soils. In: Peter K, editor. Horticulture for Nutrition Security. New Delhi: Daya Publishing House; 2015. pp. 169-188

[21] Manihuruk E, Harianto, Kusnadi N. Factors affected farmers to choose the cassava's cropping pattern and technical efficiency in Centre Lampung. Ağrı. 2018;**17**(21):139-150. DOI: 10.31186/jagrisep.17.2.139-150

[22] Ginting E, Sundari T, Triwiyono B, Triatmodjo. Identifikasi Varietas/Klon Ubikayu Unggul untuk Bahan Baku Bioetanol. Penelitian Pertanian Tanaman Pangan. 2011;**30**(117):127-136

[23] Sundari T. Pengenalan Varietas Unggul dan Teknik Budidaya Ubi kayu, no. Report No. 55.STE. Final. Malang: Balai Penelitian Kacang Kacangan dan Umbi Umbian; 2010

[24] Radjit BS, Saleh N, Subandi E Ginting E. Teknologi produksi ubi kayu mendukung industri bioetanol. Bulletin Palawija. 2008;**36**(16):27-36

[25] El-Sharkawi MA, Cadavid LF. Genetic variation within cassava germaplasm in response to potassium. Experimental Agriculture. 2000;**36**: 323-334

[26] Wargiono J. Pemupukan NPK dan Sistem Tanam Ubikayu pada Tanah Ultisol Lampung. Penelitian Pertanian Tanaman Pangan. 2003;**22**(3):114-120

[27] Ernawati R. Kajian Budidaya Ubikayu (Manihot esculenta Crantz) Sambung. Jurnal Pengkajian dan Pengembangan Teknologi Pertanian. 2010;**13**(2):85-92

[28] Najibl MF, Setiawan K, Hadi MS, Yuliadi E. Perbandingan Produksi Ubikayu (Manihot esculenta Crantz) Akibat Penambahan Pupuk KCl dan Pupuk Mikro Saat Panen 7 Bulan. Jurnal Kelitbangan. 2020;**8**(3):237-252

[29] Mossor-pietraszewska T. Effect of aluminium on plant growth and metabolism. Acta Biochimica Polonica. 2001;**48**(3):673-683

[30] Zotarelli L et al. Impact of tillage and crop rotation on aggregate-associated carbon in two Oxisols. Soil Science Society of America Journal. 2005;**69**: 482-491

[31] Nakviroj C, Paisancharoen K, Boonseng O, Wongwiwatchai C, Roongruang S. Cassava long-term fertility trials in Thailand. In: Howeler R, editor. Proceedings of the Cassava Research and Development in Asia: Exploring New Opportunities for an Ancient Crop. Proc. 7th Regional Workshop. Bangkok. 2002. pp. 212-223

[32] Howeler RH. Long-term effect of cassava cultivation on soil productivity. Field Crops Research. 1991;**26**:1-18

[33] Asnawi R, Arief RW. Teknologi Budidaya Ubikayu. In: Slameto', Kiswanto, Wijayanto B, Hermanto,

Dalmadi, editors. Balai Besar Pengkajian dan Pengembangan Teknologi Pertanian (BBP2TP). Bogor. 2008. ISBN 978-979-1415-27-9

[34] Howeler RH. Cassava mineral nutrition and fertilization. In: Hillocks R, Thresh J, Belloti A, editors. Cassava Biology, Production and Utilization. Bangkok: CAB International 2002; 2002. pp. 115-147

[35] Chen JH. The combined use of chemical and organic fertilizers and/or biofertilizer for crop growth and soil fertility. In: Proceedings of the International Workshop on Sustained Management of the Soil-Rhizosphere System for Efficient Crop Production and Fertilizer Use. Bangkok. 2006. pp. 1-11

[36] Charpentier H, Razanamparany C, Andriantsilavo M. Intercropping cassava with Brachiaria sp. on degraded hillsides in Madagascar. Congrès mondial d'agriculture de conservation. 2005;**3**: 12-17 [Online]. Available: http://publications.cirad.fr/une_notice.php?dk=536192

[37] Husson O, Tuan HD, Boyer J, Chabanne A, Caesar-Thon That TC, and Seguy L. Impacts of cropping practices and direct seeding on permanent vegetal cover (dspvc) techniques on soil biological activity in northern vietnam. In: II Congresso Mundial sobre Agricultura Conservacionista IMPACTS. 2002. Vol. 1. pp. 460–463. [Online]. Available: http://madadoc.irenala.edu.mg/documents/8076_Extended summary poster biology_Vietnam_Igacu.pdf

[38] Thierfelder C, Amèzquita C, Stahr K. Effects of nine cassava-based cropping system on superficial soil structural degradation in The Andean Hillsides of Colombia. In: 13th International Soil Conservation Organisation Conference. Conserving Soil and Water for Society: Sharing Solutions. Brisbane. 2004. pp. 1-4

[39] Lu W, Zhang F, Cao Y. Mobilization of soil phosphorus by low-molecular-weight organic acids. Plant Nutrition Development Plant Soil Science. 2001; **92**:554-555. DOI: 10.1007/0-306-47624-X_268

Section 3

Cassava Disease Control

Chapter 7

Challenges of Cassava Mosaic Begomoviruses, Cassava Brown Streak Ipomoviruses and Satellites to Cassava Production

Stephen Kwame Torkpo and Emmanuel Amponsah

Abstract

Cassava is an important food security and industrial crop. Its production is constrained by viral diseases such as cassava mosaic disease (CMD) and cassava brown streak disease (CBSD), caused by cassava mosaic begomoviruses (CMBs) and ipomoviruses, respectively. In recent years, CMBs have been associated with satellite DNAs. Food security status of cassava coupled with high demand for cassava as feed and industrial uses has been the driving force for scientists and the research community in Africa and beyond. In this review, cassava production, uses of cassava, production constraints, begomoviruses, satellite DNAs, *Bemisia tabaci,* cassava mosaic disease (CMD), *Cassava brown streak virus (*CBSV), current and future efforts in cassava production and research are discussed. This was done in an effort to create a knowledge pool that can promote cassava food security status and mitigate disease and yield loss.

Keywords: cassava, production, constraints, cassava mosaic disease, cassava brown streak disease

1. Introduction

Roots and tubers are important food crops, marketed globally with an income generation and food security role recognized by consumers and the research community [1]. Cassava (*Manihot esculenta* Crantz) is grown in the tropics (sub-Saharan Africa, Asia and Latin America). The crop has diverse uses and applications in food, green energy and feed and an important source of income for local farmers at the subsistence level [2, 3]. In sub-Saharan Africa, cassava production has been a major contributing sector with a steady increase in annual production since the last two decades [4]. In Ghana, farmers are self-sufficient in cassava production with high per capita consumption [4, 5]. In Latin America, cassava produced is utilized as food and feed mostly, with few used for industrial purposes [3, 6].

Amidst the role of cassava in food security, there is the existence of a large yield gap in the sector due to diseases, inadequate research funding, unbalanced crop nutrition, poor agronomy and soil fertility management and low level of adoption by local farmers in using tolerant cultivars [7, 8]. Cassava mosaic disease (CMD) is the most prevalent biotic limitation in sub-Saharan Africa (SSA) with newer reports of its incidence in Asia [7, 8]. Lower yield values are recorded which can also be attributed to recombinant *begomovirus* strains and the emergence of satellites that modulate and increase disease severity in SSA [8]. Occurrence of cassava brown streak in SSA further threatens cassava production. The current trends on cassava as a food security crop, cassava viruses and their effects on cassava production in the last few decades and the way forward have been discussed.

2. Cassava production

Cassava (*M. esculenta;* family *Euphorbiaceae*) is a food source for large number of people in developing countries. It is cultivated mostly in the tropics [9]. According to [10, 11], global cassava production has doubled from 124 to 277 million tons in 36 years (1980–2016) due to the ever-increasing demand for cassava for food, feed and industrial products such as biofuel and starch. Africa contributes 57% of the world's cassava production, with Nigeria contributing 21% of global production and thus considered the top global producer, followed by Asia with a production 32% and Americas about 11% [3].

In Asia, production area was estimated around 4.0 million ha [6, 12] with output of 76.6 million tons [6]. In Latin America and the Caribbean, where cassava originated, the area cultivated to cassava was 2.6 million ha with and output of 34.3 million tons [6]. In Cote d'ivoire, 6.7 t/ha productivity has been recorded, which is lower than the 9.8 t/ha average yield in Africa [13]. The ten leading cassava producing countries are presented in **Table 1**.

2.1 Cassava as a food security, feed and industrial crop

Cassava is a food security crop with characteristic high calorie yield per hectare, disease tolerance, flexible time to harvest as compared to other crops [3]. The crop is cultivated in over 40 countries worldwide and over 70% of SSA's production recorded in Nigeria, Congo DRC and Ghana [10]. The crop is grown in less fertile marginal soil conditions thus providing income for farmers in marginalized areas, and food for households [14]. The crop provides 250 kcal/ha/day of energy compared to maize, which provides 200 kcal/ha/day [7]. Cassava tuberous roots, a staple food for many across Africa provides rich carbohydrate energy source and an essential food security crop [14]. In sub-Saharan Africa, cassava is processed into producys such as attiéké (cassava couscous), gari (toasted granules), placali (paste), and futu (pounded cassava mixed with pounded plantain), and several other forms [13].

As a food crop, cassava is the most beneficial stable crop consumed by over 800 million people [3, 15]. Cassava provides over 700 million people an energy calorie of 500 cal/day of energy with a 100 cal/day consumption of the roots in tropical areas [3]. The tuber is also processed into animal feed [13]. In Asia, cassava is cultivated for human consumption with yield averaged at 34 t/h, and serves as a secondary staple crop and delicacy in households and hotels whilst 10% of production is processed into fermented flour products [3]. The crop is also consumed as dried cassava chips and biofortified commercial livestock feed for animals [6].

	Country	Production volumes							
		2013	2014	2015	2016	2017	2018	2019	2020
1	Global	275.98 M	282.63 M	282.10 M	281.77 M	275.15 M	285.01 M	293.15 M	296.22 M
2	Nigeria	47.41 M	56.33 M	57.64 M	59.57 M	55.07 M	55.87 M	59.41 M	60.00 M
3	Democratic Republic Congo	33.92 M	34.87 M	34.93 M	35.50 M	37.70 M	38.87 M	40.05 M	41.01 M
4	Thailand	30.23 M	30.02 M	32.36 M	31.16 M	30.50 M	29.37 M	31.08 M	29.00 M
5	Ghana	15.99 M	17.80 M	17.21 M	17.80 M	19.01 M	20.85 M	19.37 M	21.81 M
6	Indonesia	23.94 M	23.44 M	21.80 M	20.26 M	19.05 M	16.12 M	16.35 M	18.30 M
7	Brazil	21.48 M	23.25 M	23.06 M	21.04 M	18.50 M	17.88 M	17.59 M	18.21 M
8	Vietnam	9.76 M	10.21 M	10.74 M	10.91 M	10.27 M	9.85 M	10.17 M	10.49 M
9	Angola	16.41 M	7.64 M	7.73 M	7.92 M	8.33 M	8.73 M	9.00 M	8.78 M
10	Cambodia	7.55 M	7.50 M	7.50 M	7.50 M	7.50 M	7.50 M	7.50 M	7.66 M

(Credit https://www.tridge.com/intelligences/mandioca/production); M, million metric tons.

Table 1.
Cassava production trends of the top 10 producers of cassava from 2013 to 2020.

Over the past 10 years, increasing demand for cassava processed starch has been attributed to the high attractiveness of its allergy-free, and freeze-thaw stability properties to diverse food and non-food industries [16]. Cassava starch has 0.06–0.75% protein, 0.11–1.9% fiber, 0.03–0.29% ash, 0.0029–0.0095% phosphorous, 0.01–1.2% lipid contents, respectively [17]. Furthermore, a new trend of growing demands towards starch-based ingredients has been reported; an interesting focal point of starch processing [16]. Despite Nigeria's role as the top producer of cassava globally, Thailand tops the chart for the leading producer of cassava starch [16].

Low gelatinization and retrogradation, viscosity and higher water-binding capacity properties of cassava starch makes it preferable in food, pharmaceutical and chemical products [18]. Nevertheless, the high digestion rate of cassava starch resulting in the increased risk of cardiovascular disease has been the major drawback [16]. Cassava starch has diverse applications in the food industry with products such as noodles, tapioca pearls, sweeteners (e.g. dextrin, glucose, monosodium glutamate and high-fructose syrup), pastry products, yoghurts and microbial fermented feedstock [17]. Inclusion of products in rations increased cattle liveweight gain (LWG) and metabolizable energy content [19]. Furthermore, in the non-food industries, cassava starch is used for paper products, adhesives, pharmaceutical and textiles [16].

Cassava peels have been beneficial in the cassava utilization chain as animal feed supplement which is nevertheless not preferable [20]. Energy production is also a benefit derived from cassava peel biochemical and thermo-chemical processing [20]. Biochemical processes involve bioethanol fermentation and anaerobic digestion whereas thermochemical processes comprise pyrolysis, gasification, combustion, and liquefaction [21]. However, these processing technologies are not duly established in Africa [19, 20].

Biorefining involves the utilization of zero waste technologies to produce renewable energy [22]. Anaerobic digestion converts organic matter into biogas and biofertilizer which has led to good and controlled waste management system energy [22].

3. Production constraints

Cassava production is susceptible to several biotic constraints including pests and diseases. The most detrimental viral diseases in SSA are cassava mosaic disease (CMD) and cassava brown streak disease (CBSD) [2, 23]. Massive crop losses and reduced crop productivity over the last two decades have been worsened by the re-emergence of CMD epidemics and the emergence and evolution of new recombinant geminiviruses, the vector population increment and whitefly 'B' biotype activities [23]. CMBs affect the yield of infected crops and reduce the growth of local unimproved varieties to its susceptibility status and can result in crop failure of up to 100% yield loss [24]. CMD and CBSD cassava sources are present in the indigenous African flora, which are influenced and made worsened by factors such as susceptible varieties cultivation, abundance of transmission efficient insect vectors, and use of infected planting materials from previous harvests [24].

Cassava brown streak disease (CBSD) and cassava mosaic disease (CMD) results in economic losses estimated at over $1.0 billion per annum, adversely threatening food security in SSA [25, 26]. Crop losses attributed to CMD is characterized and documented over the years as leaf and root damages which cause food storage deficiencies in the root region [25]. CMD characteristic symptoms in infected plants (**Figure 1**) include "green to yellow" foliar mosaic, narrowing, twisting, leaf malformation

Figure 1.
CMD-affected cassava plant.

and in most severe conditions, leaf abscission and reduction, and stunted growth, which results in few or no tuber production, causing significant yield loss [27]. CMD infection results in loss of planting materials and render stem cuttings unsuitable and unclean for propagation [27]. Furthermore, CMD-affected cassava plants exhibit varied symptom expression, of which various factors contribute to these variations. These factors include the infection time, virulence of virus species, synergism, age of the crop, genome integration of the virus, the specific strain of virus, sensitivity of plant host and other biotic factors [25, 28]. In Kenya, Sudan, Tanzania, and the Democratic Republic of Congo, CMD pandemic results in a similar poor yield in cassava cultivation [25]. CMD decrease crop yields, which is worsened by susceptible cultivar cultivation across SSA by farmers [13, 25].

The widespread and aftermath of the CMD epidemic in Uganda led to highly destructive economic losses in global cassava production, complete crop abandonment by farmers, food shortage, feed and fiber loss in rural homes, and famine-related deaths [27]. Food shortages from the catastrophic epidemic exacerbated the food supply deficit of at least 800 million people having poor nutrition which compromised economic and food security [29].

Diverse CMD economic losses estimations have been documented over the years and the difference in these reports can be attributed to the location of CMD distribution and the time and year of infection; these estimations range between 20% and 95% [30]. CMD annual losses in cassava production at an estimated amount of more than US$1 billion have been reported [25, 31].

3.1 Begomoviruses

Begomoviruses are members of the plant virus family *Geminiviridae,* characterized by twin geminate icosahedral particles (22 nm × 38 nm) with circular single stranded

DNA (ssDNA) genome [23]. These viruses are the second largest plant virus family. Begomoviruses are transmitted by the whitefly, *Bemisia tabaci* Gennadius (*Hemiptera: Aleyrodidae*), and through infected stem cuttings used routinely by most local farmers [25]. The viruses cause severe damage and negative impact on cassava production in Africa [28]. Begomoviruses have monopartite (DNA A) and/or bipartite (DNA A and DNA B) circular ssDNA molecules (2.7–3.0 kb) and are the leading biotic constraint in cassava production due to its persistent re-emergence and recombination events in Africa [30, 32].

Monopartite begomoviruses are commonly found in Africa, Indian, Mediterranean and European region, and are generally classified as the Old World begomoviruses (OW) whilst bipartite begomoviruses are frequently found in South and Central America are generally classified as the New World (NW) begomoviruses [33]. Begomoviruses have different types of genomes (**Figure 2**). The Old-World DNA A encodes six proteins (CP, Rep, Ren, V2, TrAp, and C4), of which majority are associated with betasatellites (~1350 nucleotides (nt) circular ssDNA molecules) of the family *Tolecusatellitidae* and genus *Betasatellite* [34]. These betasatellites depend on a main virus for movement, plant transmission and replication [35]. The DNA-A genome consist of 6 ORFs; 4 ORFs (AC1, AC2, AC3 and AC4) on the complementary strand and 2 ORFs on the virion strand (AV1, AV2) needed in encapsidation and replication, whilst the DNA-B genome possess two ORFs (BV1 and BC1) vital for intra- and inter- cellular movement respectively.

DNA A component encodes two overlapping virion-sense ORFs (open reading frames) involved in encapsidation (AV1) and suppressor targeting PTGS silencing (AV2), replication and transcription using overlapping complementary-sense OFRs (AC1, AC3) and host mediated gene silencing suppression (AC2, AC4) [34]. On the other hand, DNA B component encodes two nonoverlapping ORFs (BV1, BC1) on the virion and complementary strands respectively for inter- and intracellular virus trafficking [32]. DNA-A and DNA-B components are responsible for systemic infection, despite DNA-A single role in disease symptom induction [36]. The leftward and rightward DNA-B and DNA-A transcriptional units are divided by a 200 nt (nucleotides) homologous intergenic region (IR) [32].

Cassava mosaic begomoviruses occur both in sub-Sahara Africa and parts of Asia (**Figure 3**). Different *Cassava mosaic begomovirus* species have been classified; *African*

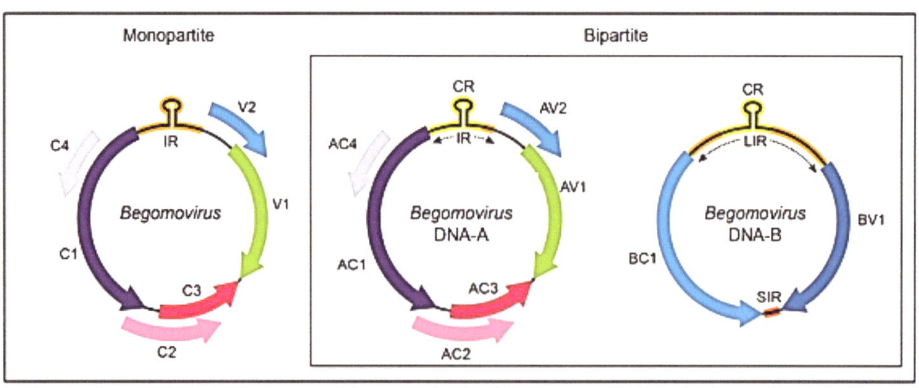

Figure 2.
Genome organization of begomoviruses. The open reading frames (ORFs) are on the complementary and virus strands (credit: ICTV).

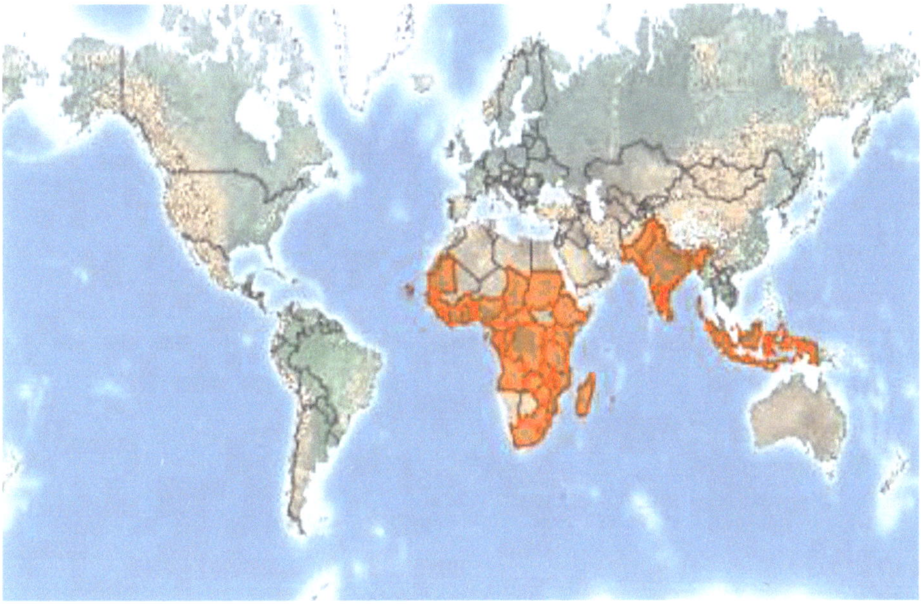

Figure 3.
Distribution of cassava mosaic begomoviruses in the world (credit: CABI).

cassava mosaic virus (ACMV), *East African cassava mosaic Cameroon virus* (EACMCV), *East African cassava mosaic virus* (EACMV), *East African cassava mosaic Zanzibar virus* (EACMZV), *East African cassava mosaic Kenya virus* (EACMKV), *South African cassava mosaic virus* (SACMV), *East African cassava mosaic Malawi virus* (EACMMV), *Indian cassava mosaic virus* (ICMV) and *Sri Lanka cassava mosaic virus* (SLCMV) [37] and *Cassava mosaic Madagascar virus* (CMMGV) [38].

Cassava mosaic begomoviruses are persistently transmitted and can occur in both single and co-infections [39]. ACMV, EACMCV and EACMV are widely prevalent and distributed in sub-Saharan Africa and a recombinant EACMV-UG strain has been reported with shared genomic properties with both ACMV (16% similarity of DNA-A genome) and EACMV (84% similarity of DNA-A genome) [9, 27]. EACMV-UG reports have been documented in SSA [27, 40]. The rapid spread of EACMV-UG can be attributed to inefficient sub-regional quarantine programmes in many countries and the indiscriminate dissemination of CMD-affected stem cuttings across sub-Saharan Africa [27]. Nevertheless, other recombinant CMBs are localized in distribution, confirming EACMV-UG better viral adaptation in sub-Saharan Africa [27].

3.2 *B. tabaci*

Whiteflies, *B. tabaci* (Hemiptera: *Aleyrodidae*) (**Figure 4**), are responsible for transmission of CMBs. CMB acquisition time is accelerated when non-viruliferous whiteflies are starved before acquisition feeding on CMD-affected cassava plants, following 6–8-hour latent period prior to transmission of virus and virus retention for about 9 days [41]. Inoculation access period of 10–30 minutes is required by viruliferous whiteflies for virus inoculation into healthy cassava plants [28]. Considerable amount of resources, and breeding efforts on *B. tabaci* and its role in CMD spread is vital for successful management of the disease.

Figure 4.
Whitefly (Bemisia tabaci) infestation on leaf tissues (credit: CABI).

3.3 Satellite DNA

DNA satellites are monopartite virus-associated subviral agents, circular ssDNAs with a conserved nonanucleotide sequence "TAATATTAC" that depend on the host cell co-infection with a helper virus for multiplication [42]. Satellites do not replicate within host cell without the helper virus and do not play any role in helper virus' life cycle [28].

DNA satellites presence in *Cassava mosaic begomovirus* infection magnifies, CMD symptom severity through constitution and host cells activities, increases viral accumulation, further worsening symptoms expression and increase CMD-induced losses which differs significantly from disease symptoms of the helper virus only [30]. Molecular characterization led to the detection of two novel small DNA molecules (satII and satIII), and confirmed these molecules as satellite DNAs due to their dependency on geminiviruses for movement and replication [43]. SatDNAs have a vital characteristic feature of breaking down resistance in resistant cultivars such as the West African cassava landrace TME3. SatDNAs have GC-rich region with direct repeats of short pentanucleotides CCGCC, trinucleotides CGC, and hexanucleotides CCGCCG, and have no origin of replication [44]. Bipartite begomoviruses associated satellites are found in replication and movement within the plant [28]. SatDNA-II has one putative TATA Binding Protein (TBP) motifs [43] and SatDNAIII have three putative TATA Binding Protein (TBP) motifs (GATATAAATA, TACATATATAT and TCTGTATATA) [28]. SatDNAs genomes have putative consensus transcription poly (A) signal AATAAA, with a positioned motif TTGTA upstream, making SatDNAs biologically functional [28, 30, 43].

DNA satellites have raised concerns pertaining to its impact on cassava production and vital role in exacerbating CMD severity caused by CMBs and substantial reduction in yield across Africa [28]. DNA satellites reduce yield as a result of virus infection, limits the distribution, exchange and multiplication of stem propagules, affecting conventional cassava breeding programmes [44]. SatDNA-II induced severe CMD symptoms expressed through mosaic, yellowing and distorted leaf symptoms and has been reported to be mostly associated with EACMV-UG whereas SatDNA-III induces unique CMD symptoms expressed through prominent yellowing and severe narrowing of leaves and mostly associated with EACMV-TZ [44].

The effectiveness of interaction between begomoviruses and satellites have common features such as the presence of stem loop for DNA replication, common genetic architecture, differentiated phloem-associated cells replication, length of individual genes at specific locus and localization [28]. In Tanzania, [28] stressed widespread occurrence of satellites (SatDNA-II and SatDNA-III) concurs with main CMB distribution and an evident of sequence integration of satellite into host cassava genome and satellites isolates mixed existence in the infected plant genome. From the study, the diversity in biological functions of SatDNA-II and SatDNA-III were confirmed [44]. In Ghana, [8] reported occurrence of SatDNAII and III for the first time, and its potential threat to the nation's food security and that of the sub-region cannot be overemphasized.

3.4 Cassava mosaic disease (CMD)

CMD is widespread and has been reported in major cassava growing regions of SSA. Severe CMD transmission is influenced by the flight and feeding activities of superabundant *B. tabaci* populations at 20 to 30 km/year [45].

Cassava mosaic disease has three distinguishing disease situations (epidemic, endemic and benign) as assessed in relation to pests and diseases; with rapid spread and transmission in epidemic situations causing severe and prevalent impact as reported in Uganda in 1990s [24, 28]. *East African cassava mosaic virus*–Uganda (EACMV-UG), a co-infected recombinant virus formed through a synergistic interaction of EACMV and ACMV caused the CMD epidemic in Uganda in the late 1980s and early 1990s [45]. Endemic areas were characterized with high incidence of CMD but relatively low symptoms expression whereas the benign situation is characterized with below 20% (low) CMD incidence [28, 43].

Three decades of the severe CMD pandemic in Uganda has passed and the devastating widespread of CMD continuous advancements to Democratic Republic of Congo in southern Africa and West African countries such as Nigeria and Ghana [8, 13, 45]. The CMD pandemic continues to pose a threat to cassava production in sub-Saharan Africa [45].

There have been reports from various countries in sub-Saharan Africa confirming CMD occurrence across the continent [36]. These reports established and revealed that CMD severity is influenced by CMBs mixed infection and synergistic effects with associated DNA satellites [44]. A nationwide survey by [29] confirmed a new recombinant strain of EACMV and six CMG species with novel begomoviruses; revealing increased CMD geographical distribution and diversity in Kenya. Also, CMD is endemic in the northern and coastal regions of Ivory Coast [13].

Reports have confirmed the distribution of the CMD viruses in a dynamic fashion; EACMV and ACMV is restricted to "non-overlapping geographical areas" in southern Africa, whereas EACMV only occurred in the Mozambique, Malawi, Zimbabwe and Madagascar [46]. Recent reports by [2, 36] re-confirmed EACMVs in Togo, Western Kenya, Nigeria, North-Eastern Zambia, Ivory Coast, Western Tanzania, Cameroon, Democratic Republic of Congo and Guinea.

Molecular detection (using Enzyme linked immunosorbent assay and Polymerase chain reaction) and CMB distribution maps revealed the presence of ACMV in all regions across SSA; from the northern regions of South to the savannah zones of Sahel [27, 36]. In Madagascar, a strong cohesive evolutionary interaction in CMB species, coupled with EACMV-like lineages in the archipelagos of Seychelles and Comoros which was transmitted to these islands three decades ago fits through recombinant of CMBs and abundant whitefly population [23].

3.5 Cassava brown streak virus (CBSV)

CBSVs belong to genus, *Ipomovirus* and family *Potyviridae*. *Ipomoviruses* possess single-stranded genomes, positive sense in nature and have large polyproteins and ten mature proteins [47]. Molecular characterization of coat protein sequences confirmed and established two genetically distinct species: *Ugandan Cassava brown streak virus* (UCBSV; 87–90% amino acid identity) and *Cassava brown streak virus* (CBSV; 76–78% nucleotide). CBSV and UCBSV are African indigenous ipomoviruses and occur in Africa only; whitefly vectors are the transmitting vectors [48].

CBSVs share unusual features when analyzed at the genomic level and do not possess the multi-functional helper-component proteinase protein (HCPro) (**Figure 5**), which is present in other *Potyviridae* viruses except for *Squash vein yellowing virus* (SqVYV) and *Cucumber vein yellowing virus* (CVYV) [47, 49]. The absence of HCPro is replaced by P1 serine proteinase's silencing suppressor activity [49]. The P1 proteins of UCBSV and CBSV contain LXKA motifs and zinc finger [50]. Occurrence of CBSV in infected cassava in East Africa has been reported. Recent whole genome sequencing and phylogenetic analysis proved the existence of three new species in the clade of UCBSV and the non-limitation of the viral species to agro-ecological zones [51].

The crop cultivars respond diversely to CBSD through range of symptoms in sensitive cultivars and foliar symptoms coupled with mild root necrosis in tolerant cultivars at different infection time [50, 52]. NASE 3 cultivar remained susceptible to

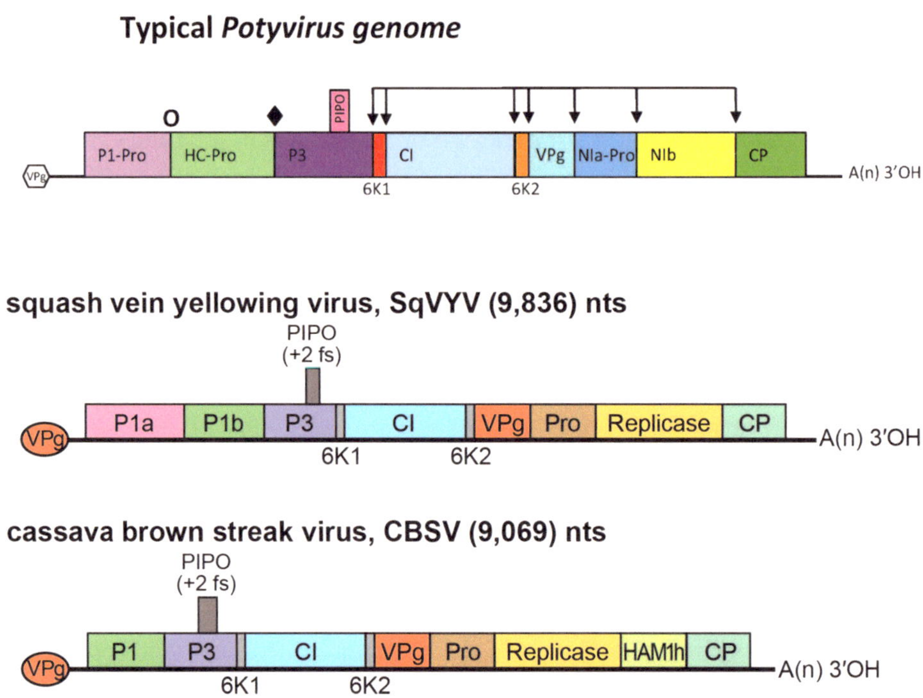

Figure 5.
Genomic organization of Squash vein yellowing virus (SqVYV) and Cassava brown streak virus in comparison with a typical Potyvirus genome (Credit: ICTV).

CBSV but exhibits high levels of resistance to UCBSV infection [53]. However, [54] reported that tolerant cultivars withstood lower viral titres as compared to susceptible cultivars. No correlation between symptom severity and viral load has been established; NASE 14 cultivar expresses severe root necrosis even with low viral accumulation whereas NASE1 cultivar expresses no foliar or root necrosis symptoms despite having a high viral accumulation [52].

3.6 Cassava brown streak disease (CSBD)

CBSD was detected initially in infected cassava crops in Tanzania in 1936; and in inland Uganda and Malawi in 1950 [50]. Aftermath of CSBD ignorance and undue lack of attention over several decades resulted in its re-emergence as an epidemic in the 1990s that devastated cassava production and a shocking threat to food security [55]. The sudden occurrence was reported initially in the East African region, and spread to other countries in the region [55]. By 2010, CBSD had translated into a regional pandemic in Central and East Africa; with a strong spread signal to other sub-Saharan countries, posing serious menace to tuber yield [56, 57].

The disease results in infected crops expressing symptomatic brown-black, necrotic root rot and constriction, brown stem streaking and adversely diminishes tuber root yields [55]. For CBSD varying foliar symptoms, chlorotic mottles, blotching, leaf vein chlorosis are eminent but very subtle for morphological detection and even asymptomatic in certain reports [55]. As such, asymptomatic infected cuttings are distributed and transplanted by local farmers, increasing the disease spread [56]. The whitefly *B. tabaci* semi-persistently transmits CBSV and UCBSV [58]. CBSD symptoms vary and is influenced by the biotic conditions present, viral strain present, cassava cultivar, severity, crop part affected, age of the cassava crop during infection time, and onset of symptom expression [50]. Foliage and root symptoms expressed by the two causal agents differ; UCBSV elicits circular leaf vein chlorotic blotches, whilst CBSV elicits feathery chlorosis of the vein, blotches, developed chlorotic and severe root necrosis [50]. CBSD-affected cassava roots showing necrosis symptom are presented in **Figure 6**.

No recombination event exists currently but co-infection synergy has been reported which may lead to severe symptoms expression [55]. The CBSD pandemic disease are a threat to food security in sub-Saharan African and are further exacerbated by the distribution of CMD-resistant cassava that are asymptomatic for CBSD, CBSD-infected cassava propagule transportation without the necessary inter-regional phytosanitary measures and frequency and abundance in polyphagous whitefly vectors [48, 59].

Similarly, confirmed recent re-emergence of CBSD has been made in many East African countries such as Democratic Republic of Congo [60], Burundi [61] and South Sudan have been documented [50].

3.7 Economic importance of CBSD

CDSD causes detrimental damage and impact to cassava production; estimation of its impact and loss economically per annum is pegged at US$750 million across East and Central Africa [62]. CBSD is one of the devastating factors of increasing cassava yield losses in East Africa [56]. CSBD's on-going distribution worsens current food insecurity in SSA and also possible spread to West Africa [56]. The sudden attention and influx in scientific reports, reviews and community discussions on CSBD and its

Figure 6.
CBSD-affected cassava tubers (credit: CABI).

threat to food security is a topic of great concern on CSBD epidemiology [50]. CBSD expansion across Central and East Africa has heighted the sudden need for initiation and deployment of CSBD control measures [50, 56].

4. Current and future efforts in cassava production and research

Since the aftermath of CMD detection in Africa, substantial control efforts have been documented, of which phytosanitation and resistance breeding has been the most reported. Cassava breeding has resulted in the production of CMD resistant/tolerant high yielding cassava cultivars [63]. In East Africa, b-carotene (provitamin A) breeding focused objective has been implemented for the genetic improvement of cassava to produce carotenoids-rich cassava varieties which sought to promote dietary Vitamin A deficiency alleviation [64]. Cassava breeding programmes in SSA researched and exploited CMD2 locus in the deployment of highly heritable and resistant CMD genotypes [65].

Efforts to control CMD has been focused on virus-free deployment and CMD-resistant germplasm dissemination, which is restricted by long and tedious process of traditional breeding and lack of CMBs genomic characterization [27]. However, well implemented and carefully managed field research is expensive and demands more funding from stakeholders [64]. The increasing demand for food, feed and biomass-based products globally is keen in transforming SSA agriculture industry scope to a "biomass-supplying" hub [4]. The result-oriented research efforts, suitable fertilization and appropriate weed, disease and pest control, is projected to increased yields by 50–100% [66].

Amidst the efforts and research contributions over the last two decades, CMD is prevalent and cassava production has been constrained as a result of virulence of begomoviruses, co-infection synergy of CMBs, integration potential of CMBs and associated satellites into host cell, association of CMBs with satellite DNAs through reassortment and recombination events, high fecundity and spread of *B. tabaci*, varied climatic factors, and use of infected propagules by local farmers [28]. As such we cannot ignore the activities of satellite DNAs and widespread activities of recombinant virulent strains [8]. Sub-Saharan Africa needs a phytosanitary facilities and techniques equipped with affordable detection approaches.

Limited knowledge and insights on CBSD viral variations and its complex interactions, cassava cultivars, vectors, and biotic conditions influence in disease spread [50]. CBSD prevalence, incidence, distribution and whitefly populations in farmers' fields need periodic and regular monitoring to document periodic changes in disease distribution in affected regions [50, 67]. Development of predictive models with evident-based for CSBD control decisions needs to be developed, which is dependent on farmer engagement, education on disease management and information for stakeholder awareness as implemented in Uganda [50].

Lessons need to be learnt from the ignorance and delayed attention to the CSBD outbreak to prevent and minimize the impact of future outbreaks. As such the scientific community needs to put in measures for CSBD epidemiology, outbreak into new areas prediction and spread [68].

Advances in molecular approaches including genetically modified resistant lines development, single multiplex RT-PCR reaction for CMB detection, RT-PCR optimization for CBSV and UCBSV detection, certified virus-clean planting material provision, marker-assisted breeding and sensitive diagnostics utilization have been applicable and useful in the CSBD progress status [50, 69]. Deployment of next generation high through-put sequencing (NGS) has ensured virus-free planting material (95% detection rate) through screening cassava crops for CBSVs prior to planting material dissemination [50, 70]. Despite these recent progress, CMD and CBSD biology and epidemiology areas need further attention and the accessibility of affordable diagnostic techniques in sub-Saharan Africa for local detection [50].

5. Conclusion

Cassava, a food security crop has huge potential for global food production as demand for the crop increases. In the last two decades, cassava mosaic disease and cassava brown streak disease have become major threats to food security across SSA, supply of feed and raw materials to industry. As such, further research and control efforts in achieving world food security status of cassava are necessary. Development of future research programmes is necessary to increase global cassava yield. Integration of different approaches and implementation of comprehensive cassava mosaic disease and cassava brown streak disease management in the region is required.

Conflicts of interest

The authors declare no conflict of interest.

Author details

Stephen Kwame Torkpo[1,2]* and Emmanuel Amponsah[1]

1 Department of Crop Science, School of Agriculture, College of Basic and Applied Sciences, University of Ghana, Legon, Ghana

2 Forest and Horticultural Crops Research Centre, School of Agriculture, College of Basic and Applied Sciences, University of Ghana, Legon, Ghana

*Address all correspondence to: sktorkpo@ug.edu.gh

References

[1] Petsakos A, Pragerb SD, Gonzalez CE, Gamac AC, Sulserd TB, Gbegbelegbee S, et al. Understanding the consequences of changes in the production frontiers for roots, tubers and bananas. Global Food Security. Amsterdam, New York: Elsevier; 2019;**20**:180-188

[2] Aloyce RC. Development and Evaluation of Efficient Diagnostic Tools for Cassava Mosaic and Cassava Brown Streak Diseases. Dissertation submitted in partial fulfillment for the degree of Master of Science. Johannesburg Republic of South Africa: The University of Witwatersrand; 2013

[3] Byju G, Suja G. Mineral nutrition of cassava. Advances in Agronomy. 2020;**159**:169-235

[4] Poku A, Birner R, Gupta S. Is Africa ready to develop a competitive bio economy? The case of the cassava value web in Ghana. Journal of Cleaner Production. 2018;**200**:134-147. DOI: 10.1016/j.jclepro.2018.07.290

[5] Angelucci F. Analysis of incentives and disincentives for cassava in Ghana. In: Technical Notes Series, Monitoring African Food and Agricultural Policies Project (MAFAP). Rome: FAO; 2013 © FAO 2013

[6] Howeler R, Lutaladio N, Thomas G. Save and Grow: Cassava. A Guide to Sustainable Production Intensification. Rome, Italy: FAO; 2013

[7] El-Sharkawy M. Stress-tolerant cassava: The role of integrative ecophysiology-breeding research in crop improvement. Open Journal of Soil Science. 2012;**2**(2):162-186. DOI: 10.4236/ojss.2012.22022

[8] Torkpo SK, Amponsah E, Arhin CD, Offei SK. Occurrence of cassava mosaic begomovirus-associated satellites on cassava in Ghana. Cogent Food & Agriculture. 2021;**7**(1):1963929

[9] Patil BL, Fauquet CM. Cassava mosaic geminiviruses: Actual knowledge and perspectives. Molecular Plant Patholology. 2009;**10**:685-701

[10] FAOSTAT. Production; Cassava; all Countries; 1961-2016 (Online). Food and Agriculture Organization of the United Nations. 2018 http://www.fao.org/faostat/en/#data/QC

[11] Sabitha S, Byju G, Sreekumar J. Projected changes in mean temperature and total precipitation and climate suitability of cassava (Manihot esculenta) in major growing environments of India. Indian Journal of Agricultural Sciences. 2016;**86**(5):647-653

[12] Malik AI, Kongsil P, Nguyễn VA, Ou W, Srean P, López-Lavalle LAB, et al. Cassava breeding and agronomy in Asia: 50 years of history and future directions. Breeding Science. 2020;**70**(2):1-22. DOI: 10.1270/jsbbs.18180

[13] Toualy MNY, Akinbade AS, Koutoua S, Atta Diallo H, Kumar PL. Incidence and distribution of cassava mosaic begomoviruses in Côte d'Ivoire. Research paper. International Journal of Agronomy and Agricultural Research (IJAAR). 2014;**4**(6):131-139. ISSN: 2223-7054

[14] Githunguri CM, Njiru EN. Role of cassava and Sweetpotato in mitigating drought in semi-arid Makueni County in Kenya. In: African Handbook of Climate Change Adaptation. 2021. pp. 1-19. DOI: 10.1007/978-3-030-42091-8_11-1

[15] Nassar N, Vizzotto CS, Schwartz CA, Pires OR Jr. Cassava diversity in Brazil:

The case of carotenoid-rich landraces. Genetics and Molecular Research. 2007;6(1):116-121

[16] Wang Z, Mhaske P, Farahnaky A, Kasapis S, Majzoobi M. Cassava starch: Chemical modification and its impact on functional properties and digestibility, a review. Food Hydrocolloids. 2022;**129**:107542. DOI: 10.1016/j.foodhyd.2022.107542

[17] Zhu F. Composition, structure, physicochemical properties, and modifications of cassava starch. Carbohydrate polymers. 2015;**122**:456-480

[18] Li X, Fan S, Hu W, Liu G, Wei Y, He C, et al. Two cassava basic leucine zipper (bZIP) transcription factors (MebZIP3 and MebZIP5) confer disease resistance against cassava bacterial blight. Frontiers in plant science. 2017;**8**:2110

[19] Retnaningrum S, Kusmartono K, Mashudi K, Harper J, Poppi DP. Formulating rations with cassava meal to promote high live weight gain in crossbred Limousin bulls. Animal. 2021;**15**(2):100125, ISSN 1751-7311. DOI: 10.1016/j.animal.2020.100125

[20] Kayiwa R, Kasedde H, Lubwama M, Kirabira JB. Characterization and preleaching effect on the peels of predominant cassava varieties in Uganda for production of activated carbon. Current Research in Green and Sustainable Chemistry. 2021;**4**:100083. DOI: 10.1016/j.crgsc.2021.100083

[21] Adekunle A, Gariepy Y, Lyew D, Raghavan V. Energy recovery from cassava peels in a single-chamber microbial fuel cell. Energy Sources, Part A: Recovery, Utilization, and Environmental Effects. 2016;**38**(17):2495-2502

[22] Cruz AI, Santos Andrade LR, Bharagava RN, Nadda AK, Bilal M, Figueiredo RT, et al. Valorization of cassava residues for biogas production in Brazil based on the circular economy: An updated and comprehensive review. Cleaner Engineering and Technology. 2021. DOI: 10.1016/j.clet.2021.100196

[23] De Bruyn A, Villemot J, Lefeuvre P, Villar E, Hoareau M, Harimalala M, et al. East African cassava mosaic-like viruses from Africa to Indian ocean islands: Molecular diversity, evolutionary history and geographical dissemination of a bipartite begomovirus. BMC Evolutionary Biology. 2012;**12**:228. DOI: 10.1186/1471-2148-12-228

[24] Thresh JM, Otim-Nape GW, Legg JP, Fargette D. African cassava mosaic virus disease: The magnitude of the problem. African journal of Root and tuber crops. 1997;**2**(1/2):13-19

[25] Legg JP, Kumar PL, Makeshkumar T, Tripathi L, Ferguson M, Kanju E, et al. Cassava virus diseases: Biology, epidemiology, and management. Advances in Virus Research. Amsterdam, New York: Elsevier; 2015;**91**:85-142. DOI: 10.1016/bs.aivir.2014.10.001

[26] Patil BL, Bagewadi B, Yadav JS, Fauquet CM. Mapping and identification of cassava mosaic geminivirus DNA-A and DNA-B genome sequences for efficient siRNA expression and RNAi based virus resistance by transient agro-infiltration studies. Virus Research. 2016;**213**:109-115. DOI: 10.1016/j.virusres.2015.11. 011

[27] Were MN, Mukoye B, Osogo AK, Mangeni BC, Nyamwamu PA, Ogemah VK, et al. Occurrence and distribution of Begomoviruses infecting cassava in Western Kenya. Plant. 2016;**4**(6):108-113. DOI: 10.11648/j.plant.20160406.18

[28] Ndomba O. Influence of Satellite DNA Molecules on Severity of Cassava Begomoviruses and the Breakdown of Resistance to Cassava Mosaic Disease in Tanzania. A Thesis Submitted to the Faculty of Science. Johannesburg, in fulfillment of the requirements for the degree of Doctor of Philosophy Johannesburg: University of the Witwatersrand; 2012

[29] Bull SE, Briddon RW, Sserubombwe WS, Ngugi K, Markham PG, Stanley J. Genetic diversity and phylogeography of cassava mosaic viruses in Kenya. Journal of General Virology. 2006;*87*(10):3053-3065

[30] Mascia T, Gallitelli D. CHAPTER 51. Viroids and Satellites. Bari, Italy: University of Bari Aldo Moro; 2017. DOI: 10.1016/B978-0-12-801498-1.00051-6

[31] Legg JP, Álvarez E. Diseases affecting cassava. In: Hershey C, editor. Achieving Sustainable Cultivation of Cassava Volume 2: Genetics, Breeding, Pests and Diseases. Burleigh Dodds. Vol. 10. (Burleigh Dodds Series in Agricultural Science). ISBN 9781786760043. Burleigh Dodds, Part 3; 2017. p. 322

[32] Hanley-Bowdoin L, Bejarano ER, Robertson D, Mansoor S. Geminiviruses: Masters at redirecting and reprogramming plant processes. Nature Reviews Microbiology. 2013;**11**(11):777-788

[33] Zerbini FM, Briddon RW, Idris A, Martin DP, Moriones E, Navas-Castillo J, et al. ICTV virus taxonomy profile: *Geminiviridae*. Journal of General Virology. 2017;**98**:131-133

[34] Zhang J, Dong J, Xu Y, Wu J. V2 protein encoded by tomato yellow leaf curl China virus is an RNA silencing suppressor. Virus Research. 2012;**163**:51-58

[35] Saunders K, Briddon RW, Stanley J. Replication promiscuity of DNA-β satellites associated with monopartite begomoviruses; deletion mutagenesis of the ageratum yellow vein virus DNA- β satellite localizes sequences involved in replication. Journal of General Virology. 2008;**89**:3165-3172. DOI: 10.1099/vir.0.2008/003848-0

[36] Bisimwa E, Walangululu J, Bragard C. Cassava mosaic disease yield loss assessment under various altitude agroecosystems in the Sud-Kivu region, Democratic Republic of Congo. Tropicultura. 2015;**33**(2):101-110

[37] Tiendrébéogo F, Lefeuvre P, Hoareau M, Harimalala MA, De Bruyn A, Villemot J, et al. Evolution of African cassava mosaic virus by recombination between bipartite and monopartite begomoviruses. Virology Journal. 2012;*9*(1):1-7

[38] Harimalala M, Lefeuvre P, De Bruyn A, Tiendrébéogo F, Hoareau M, Villemot J, et al. A novel cassava-infecting begomovirus from Madagascar: Cassava mosaic Madagascar virus. Archives of virology. 2012;**157**(10):2027-2030

[39] Berry S, Rey MEC. Molecular evidence for diverse populations of cassava-infecting begomoviruses in southern Africa. Archives of Virology. 2001;**146**:1795-1802

[40] Harrison BD, Zhou X, Otim-Nape GW, Liu Y, Robinson DJ. Role of a novel type of double infection in the geminivirus-induced epidemic of severe cassava mosaic in Uganda. Annals of Applied Biology. 1997;**131**:437448

[41] Dubern J. Transmission of African cassava mosaic geminivirus by the

whitefly (*Bemisia tabaci*). Tropical Science. 1994;**34**:82-91

[42] Nawaz-ul-Rehman MS, Fauquet CM. Evolution of geminiviruses and their satellites. FEBS Letters. 2009;**583**: 1825-1832

[43] Ndunguru J, Legg JP, Aveling TAS, Thompson G, Fauquet CM. Molecular biodiversity of cassava begomoviruses in Tanzania: Evolution of cassava geminiviruses in Africa and evidence for East Africa being a center of diversity of cassava geminiviruses. Virology Journal. 2005;**2**:21

[44] Ndunguru J, Fofana B, Legg JP, Chellappan P, Taylor N, Aveling T, et al. Two novel satellite DNAs associated with bipartite cassava mosaic begomoviruses enhancing symptoms and capable of breaking high virus resistance in a cassava landrace. In: Book of Abstracts, Global Cassava Partnership–First Scientific Meeting: Cassava Meeting the Challenges of the New Millennium. Ghent, Belgium: Ghent University; 2008

[45] Szyniszewska AM, Busungu C, Boni SB, Shirima R, Bouwmeester H, Legg JP. Spatial analysis of temporal changes in the pandemic of severe cassava mosaic disease in Northwestern Tanzania. Phytopathology. 2017;**107**(10):1229-1242

[46] Harrison BD, Swanson MM, Robinson DJ. Cassava viruses in the Old World. In: Proceedings: Second International Scientific Meeting of the Cassava Biotechnology Network. Bogor, Indonesia: CBN/CRJFC/AARD/CIAT; 22-26 August 1994, Working Document 150; 1995. pp. 463-472.

[47] Valli A, López-Moya JJ, García JA. Potyviridae. In: ELS. Chichester: John Wiley; 2015. pp. 1-10

[48] Jeremiah SC, Ndyetabula IL, Mkamilo GS, Haji S, Muhanna MM, Chuwa C, et al. The dynamics and environmental influence on interactions between cassava brown streak disease and the whitefly, *Bemisia tabaci*. Phytopathology. 2015;**105**:646-655

[49] Mbanzibwa DR, Tian Y, Mukasa SB, Valkonen JP. *Cassava brown streak virus* (*Potyviridae*) encodes a putative Maf/HAM1 pyrophosphatase implicated in reduction of mutations and a P1 proteinase that suppresses RNA silencing but contains no HC-pro. Journal of Virology. 2009;**83**:6934-6940

[50] Tomlinson KR, Bailey AM, Alicai T, Seal S, Foster GD. Cassava brown streak disease: Historical timeline, current knowledge and future prospects. Molecular Plant Pathology. 2018;**19**(5):1282-1294

[51] Ndunguru J, Sseruwagi P, Tairo F, Stomeo F, Maina S, Djinkeng A, et al. Analyses of twelve new whole genome sequences of *cassava brown streak viruses* and *Ugandan cassava brown streak viruses* from East Africa: Diversity, supercomputing and evidence for further speciation. PLoS One. 2015;**10**:e0139321

[52] Kaweesi T, Kawuki R, Kyaligonza V, Baguma Y, Tusiime G, Ferguson ME. Field evaluation of selected cassava genotypes for cassava brown streak disease based on symptom expression and virus load. Virology Journal. 2014;**11**:216. DOI: 10.1186/s12985-014-0216-x

[53] Ogwok E, Ilyas M, Alicai T, Rey MEC, Taylor NJ. Comparative analysis of virus-derived small RNAs within cassava (*Manihot esculenta* Crantz) infected with *cassava brown streak viruses*. Virus Research. 2016;**215**:1-11

[54] Maruthi MN, Bouvaine S, Tufan HA, Mohammed IU, Hillocks RJ.

Transcriptional response of virus-infected cassava and identification of putative sources of resistance for cassava brown streak disease. PLoS One. 2014;**9**:e96642

[55] Jones RA. Disease pandemics and major epidemics arising from new encounters between indigenous viruses and introduced crops. Viruses. 2020;**12**(12):1388. DOI: 10.3390/v12121388

[56] Legg JP, Somado EA, Barker I, Beach L, Ceballos H, Cuellar W, et al. A global alliance declaring war on cassava viruses in Africa. Food Security. 2014;**6**:231-248

[57] Shirima RR, Maeda DG, Kanju E, Ceasar G, Tibazarwa FI, Legg JP. Absolute quantification of *cassava brown streak virus* mRNA by real-time qPCR. Journal of Virological Methods. 2017;**245**:5-13

[58] Maruthi MN, Jeremiah SC, Mohammed IU, Legg JP. The role of the whitefly, Bemisia tabaci (Gennadius), and farmer practices in the spread of cassava brown streak ipomoviruses. Journal of Phytopathology. 2017;**165**:707-717. DOI: 10.1111/jph.12609

[59] Rey C, Vanderschuren H. Cassava mosaic and brown streak diseases: Current perspectives and beyond. Annual Review of Virology. 2017;**4**:429-452

[60] Mulimbi W, Phemba X, Assumani B, Kasereka P, Muyisa S, Ugentho H, et al. First report of *Ugandan cassava brown streak virus* on cassava in Democratic Republic of Congo. New Disease Reports. 2012;**26**:11

[61] Bigirimana S, Barumbanze P, Ndayihanzamaso P, Shirima R, Legg JP. First report of cassava brown streak disease and associated *Ugandan cassava brown streak virus* in Burundi. New Disease Report. 2011;**24**:26

[62] Hillocks RJ, Maruthi MN. Post-harvest impact of cassava brown streak disease in four countries in eastern Africa. Food Chain. 2015;**5**:116-122

[63] Okogbenin E, Egesi CN, Olasanmi B, Ogundapo O, Kahya S, Hurtado P, et al. Molecular marker analysis and validation of resistance to cassava mosaic disease in elite cassava genotypes in Nigeria. Crop Science. 2012;**52**(6):2576-2586

[64] Atwijukire E, Hawumba JF, Baguma Y, Wembabazi E, Esuma W, Kawuki RS, et al. Starch quality traits of improved provitamin a cassava (*Manihot esculenta* Crantz). Heliyon. 2019;**5**(2):e01215. DOI: 10.1016/j.heliyon.2019.e01215

[65] Rabbi IY, Hamblin MT, Kumar PL, Gedil MA, Ikpan AS, Jannink JL, et al. High-resolution mapping of resistance to cassava mosaic geminiviruses in cassava using genotyping-by-sequencing and its implications for breeding. Virus Research. 2014;**186**:87-96

[66] Cock JH, Connor DJ. Chapter 19 - cassava. In: Sadras VO, Calderini DF, editors. Crop Physiology Case Histories for Major Crops. Amsterdam, New York: Elsevier; 2021. pp. 588-633, ISBN 9780128191941. DOI: 10.1016/B978-0-12-819194-1.00019-0

[67] Kumakech A, Acipa A, Tumwine V, Maiteki GA. Knowledge on cassava disease management: The case of cassava brown streak disease awareness in northern Uganda. African Journal of Plant Science. 2013;7:597-601

[68] Alicai T, Ndunguru J, Sseruwagi P, Tairo F, Okao-Okuja G, Nanvubya R, et al. Characterization by next generation

sequencing reveals the molecular mechanisms driving the faster evolutionary rate of *cassava brown streak virus* compared with *Ugandan cassava brown streak virus*. Scientific Reports. 2016;**6**:36164

[69] Abarshi MM, Mohammed IU, Jeremiah SC, Legg JP, Kumar PL, Hillocks RJ, et al. Multiplex RT-PCR assays for the simultaneous detection of both RNA and DNA viruses infecting cassava and the common occurrence of mixed infections by two cassava viruses in East Africa. Journal of Virological Methods. 2012;**179**:176-184

[70] Adams IP, Abidrabo P, Miano DW, Alicai T, Kinyua ZM, Clarke J, et al. High throughput real-time RT-PCR assays for specific detection of cassava brown streak disease causal viruses, and their application to testing of planting material. Plant Pathololology. 2013;**62**(1):233-242

Section 4

Processing Cassava for Food, Feed, and Industry

Chapter 8

Cassava and Microalgae Use in the Food Industry: Challenges and Prospects

Ardiba Rakhmi Sefrienda, Dedy Kurnianto, Jasmadi Jasmadi and Andri Frediansyah

Abstract

Cassava is a good source of carbohydrates and a staple diet in many countries. It has a high-calorie count but a low protein and fat content. Microalgae biomass is increasingly being used in the food business industry due to its ease of production, low carbon requirements, and small footprint. The usage of microalgae in combination with cassava is becoming more common as it can boost the amount of nutrients in processed cassava products. In this chapter, we discuss the development of cassava products that combine cassava with microalgae. Furthermore, cassava waste contains carbohydrates, which can be used as a carbon source for the development of microalgae. Cassava starch, when modified to become cationic cassava starch, has the potential to be used as a flocculant agent for the separation of microalgal biomass. Cassava starch is also well-known for being a low-cost source of bioplastics. This chapter also addresses the possibilities for microalgae and cassava to be used as bioplastics in the same way.

Keywords: cassava, microalgae, cationic, food, bioplastics

1. Introduction

Cassava is cultivated in more than one hundred countries and is a primary source of nutrition for millions of people living in tropical areas of Africa, Asia, and America. The average amount of cassava that can be harvested from one hectare in Nigeria is 10.6 tons, making the country the largest cassava producer in the world [1]. Cassava, in addition, is considered to be one of the staple foods in Indonesia. After Nigeria and Thailand, Indonesia has emerged as one of the world's leading producers of cassava, making it one of the top three countries in this regard. Up to 53% of cassava production is used for food items, with the remaining amount being used for animal feed and as sources of bioethanol. Cassava-based food items such as boiled or fried cassava, chips, fermented cassava (*tape, peyeum*), *gatho*t, and *tiwul* are examples of these.

About 70% of the cassava root is water. If the roots are not treated within 2–3 days, they will fully oxidize. Therefore, the drying procedure is a necessary post-harvest treatment for preserving the quality of the roots. Roots of cassava that have been peeled, then cut into chips, and then dried in the sunlight. Cassava flour can be produced by grinding dried cassava chips in a mill. The locals of Java, Indonesia refer to it as "gaplek" flour. In addition to drying the roots directly, we can alter the cassava by soaking the chips in water containing bio-starter after it has been chopped. This is done after the root has been dried. There will be a three-day period of fermentation. After being fermented, the cassava chips were exposed to the sun to finish drying. In Indonesia, this type of flour is referred to as "mocaf." The acronym mocaf stands for "modified cassava flour," which is the full meaning of the term "mocaf" [2]. Mocaf flour is already being put to use as a gluten-free alternative to regular flour. It can be used as a wheat flour replacement in gluten-free baking, as well as in the preparation of gluten-free noodles and snacks. The non-food industry relies heavily on cassava flour and starch as its primary raw materials. Paper, textiles, plywood, glue, and biofuels are just some of the products that can be made with them. The gelatinization temperature of cassava starch is lower than that of other types of starch, and it also has a higher water-binding capacity and viscosity than other types of starch. These characteristics make it valuable in the culinary, chemical, and pharmaceutical industries [3]. These days, one of the components utilized in the production of biodegradable packaging is cassava starch, which is obtained from the root of the cassava plant [4].

Cassava roots have a significant amount of dietary fiber, which helps to improve heart health and eliminate atherosclerosis and the associated concerns, such as heart attacks and strokes. Cassava is also beneficial to digestive health. It contains vitamins and minerals such as vitamins C and K, as well as potassium, calcium, iron, copper, and zinc [5]. Despite the fact that cassava roots are a source of carbs, they are quite low in protein and fat content [6]. To meet the nutritional requirements for consumption, cassava root products must be coupled with foods that provide a source of protein. A protein source derived from microalgae is one of the forms of protein that are compatible with cassava food products and can be integrated with them. Microalga *Spirulina* sp. is a widespread type of microalgae that is frequently combined with foods that people consume on a regular basis, such as cookies, chocolates, energy drinks, crackers, and instant noodles [7–10]. The cassava cake and the cassava doughnut are the two cassava items that have been reported to combine with *Spirulina* sp. [11, 12]. After the addition of *Spirulina* sp., all the research revealed an increase in the product's protein content [11, 12].

The relationship between cassava and microalgae goes beyond food products. Cassava starch that has been modified to make it cationic cassava starch can be used effectively as a flocculant agent in the microalgae harvesting process [13]. Waste products from cassava processing can be used for the cultivation of microalgae seeds or for fermentation. Achi, et al. [14] reported that waste products from the processing of cassava are a significant source of pollution. Cassava peels are a significant source of waste, and in most areas, 91% of them are piled up in trash landfills [15]. Additionally, Zhang et al. [16] found that the physical and chemical properties of waste cassava were comparable to those of biomass derived from woody plants. This made waste cassava an interesting option for bioconversion into products with additional value [17]. These results imply that the waste generated during cassava processing is a significant contributor to environmental pollution and that cassava waste has the potential to be converted into value-added products. However, additional research is needed to determine whether it is possible to generate value-added commodities from cassava waste.

One example of a sustainable cycle is the integration of the cassava industry with microalgae biorefinery. It could reduce the amount of cassava wasted and the cost of microalgae biorefinery. Microalgae as a protein source has the potential to be a staple dietary fortification against malnutrition. There is also the prospect of developing cassava-microalgae bioplastics as a long-lasting, functional, and environmentally beneficial packaging material.

2. Foods derived from cassava and microalgae

Cassava production in Indonesia can reach 20 million tons per year, allowing cassava to become one of the staple foods in Indonesia. Cassava, on the other hand, is unpopular as a food source due to its stigma as a meal for marginalized people [2]. In order to increase consumer acceptance of cassava, innovation in the processing of products based on cassava is required. There have been a number of advancements made in the processing of cassava flour into food products. These include the modification of the structure of the cassava flour through the use of fermentation, as well as the transformation of well-known dishes such as cookies, cheese sticks, and noodles using cassava flour [18–21].

Cassava is a source of carbs and contains a negligible amount of protein [6]. The nutritional value of processed food products made from cassava can be improved by the incorporation of various protein sources into their composition. Microalgae are a potential source of protein that might be added to this cassava product after it has been processed. *Spirulina* sp. is by far the most common type of microalgae used.

- Product using cassava flour

Cassava flour has been used to replace wheat flour in doughnut recipes. The microalga *S. platensis* has been utilized to boost the nutritional value of cassava doughnuts. The compositions of the formulation's proximal, sensory, and technical components were assessed. When *S. platensis* was introduced to doughnuts, the protein, lipid, and fiber content increased linearly by 2.59, 3, 4, 5, and 5.41% (w/w). It also increases shearing force, making the doughnut more resistant to deformation and more difficult to chew. However, the doughnut with the highest *S. platensis* inclusion is still acceptable by the consumer, according to the acceptance test. As a result, this product can be utilized to provide improved nutrition to patients suffering from celiac disease [12].

Cassava flour can also be used to make cassava cake. Microalga *S. platensis* was introduced to compensate for the absence of protein in cassava products. The addition of *S. platensis* at 1% and 2% boosted the protein content without affecting consumer acceptance [11].

- Product using modified cassava flour (mocaf)

Food products made from modified cassava flour are becoming increasingly popular in Indonesia. The usage of mocaf can replace the use of wheat flour, which is still imported. Furthermore, an increasing number of people are concerned about gluten-free foods. The fermentation procedure used to create modified cassava flour influenced the physical quality but not the protein content. The protein content of mocaf and cassava flour is 1.2% [22]. The use of *Spirulina* sp. is one approach to boosting the protein content of these products. Commercially available mocaf and *Spirulina*-based products include 'Nasamie' mocaf *Spirulina* noodles and "Garmil" mocaf sticks as shown in **Figure 1**. In Nasamie, the addition of *Spirulina* sp. to mocaf noodles

Figure 1.
From left to the right: Mocaf noodles (https://nasamie.co.id/En), pie [22], and stick enriched with Spirulina *sp. (https://albitec.co.id/product/garmil-dari-stick-spirulina/).*

could elevate the protein level up to 8 g in an 80 g serving size (data from 'Nasamie's nutrition fact'). Therefore, in Garmil, the protein content is 3 g per 25 g serving size (data from 'Garmil Spirulina' nutrition facts) after the addition of *Spirulina* sp. to a mocaf-based stick. Furthermore, the protein content of Garmil mocaf *Spirulina* sp. sticks was higher than that of mocaf sticks reported by Kusumaningrum, Miftakhussolikhah, Herawati, Susanto, and Ariani [18]. This indicates that the addition of *Spirulina* sp. improves the nutritional quality of mocaf sticks. Al-Baarri, et al. [23] reported that the combination of 2% *Spirulina* sp. and 5% basil leaf can reduce the level of mocaf noodle hardness by as much as 50%.

Pie susu as shown in **Figure 1**, a Balinese culinary icon, is another example. It was a short pastry dough with milk and topping. Pastry dough made with a 1:2 ratio of flour and mocaf mocaf and flour. *Spirulina* sp. addition is 0.5% of the total dough. It is the maximum amount of *Spirulina* sp. that can meet the pastry requirement [22].

- Product using cassava starch (tapioca)

Tapioca is widely used in Indonesian cuisine, particularly in a variety of dishes. *Pempek, cireng, cimol, bika ambon, ongol-ongol, kue lapis*, and *cenil* are just some of the delectable dishes that are traditionally prepared in Indonesia with tapioca as one of the main ingredients. Tapioca pearls are among the most well-known foods that are generated from tapioca, and their use can be found all over the world. Tapioca in the form of balls, with a diameter of 2–8 mm [6]. Tapioca is commonly used as a topping in desserts and beverages. Frutea, a Colombian bubble tea maker, has launched Espirulitea, a bubble tea variety. Tea, milk, tapioca, and *Spirulina* sp. are among the ingredients. A franchise café in the United States created a dessert with phycocyanin extract from *Spirulina* sp., also known as blue spirulina. They create a sorbet out of coconut, pineapple, and blue spirulina. Toppings included strawberry tapioca pearls, crunchy honey GF granola, banana, strawberries, and kiwi (**Figure 2**).

Cassava products are grown in over 100 countries and feed millions of people in Africa, Asia, and some part of America. Nigeria is the world's largest cassava producer, with an average yield per hectare of 10.6 tons per acre [1]. The top three countries in terms of cassava production are Nigeria, Thailand, and Indonesia [24]. However, because of its low protein content, cassava necessitates the development of new methods of processing in order to improve its nutritional value and make it suitable for consumption as a food source in the fight against stunting. In order to

Figure 2.
Tapioca pearls using as drink and dessert enriched with spirulina sp. (https://frutea.com.co/p/espirulitea/; https://beyondjuiceryeatery.com/boba-blue-bowl/).

boost the nutritional value of cassava, one of the supplementary substances that may be used is microalgae, which is abundant in both protein and antioxidants.

3. Bioplastics from cassava and microalgae

Currently, the world faces global plastic waste contamination. Innovation is required to reduce this pollution. In contrast, plastics derived from fossil fuels have dropped. Bioplastics were utilized as an alternative to polymers derived from fossil fuels. Bioplastics are a type of plastic that can be derived from natural materials like starches and vegetable oil. The use of plant-based bioplastics is anticipated to reduce petroleum use by 15–20% by 2025, with Asia and Europe holding the biggest market share for bioplastics [25].

Bioplastics can be categorized into three groups: those made from recycled materials, those modified from naturally occurring polymers, and those made from synthetic biobased monomers. Biopolymers can be made from a wide variety of renewable but non-biodegradable raw materials, such as bio-polyethylene (Bio-PE), biobased polyethylene terephthalate (PET), and polytrimethylene terephthalate (PTT), as well as biodegradable but non-renewable materials, such as polybutylene adipate-co-terephthalate (PBAT), polybutylene succinate (PBS), and polycaprol. Starch-based polymers are the most prevalent bioplastics polymers worldwide [26]. Indonesia has mass-produced cassava starch and tapioca, for bioplastics under the brand name Telobag as shown in **Figure 3**. A Telobag is a bag constructed from telo-cassava.

Furthermore, microalgae-based bioplastics have been developed. The production of microalgal bioplastics could involve the use of microalgal biomass, bio- or petroleum-based polymers, and additives. The alternate method depends on the

Figure 3.
Bioplastics from cassava (www.telobag.com).

Species	Production methods*	Materials	Characteristic	Publication
Spirulina platensis	Melt mixing, hot molding	Glycerol, polyvinyl anhydride, maleic anhydride	Microalgal as composites	[27]
Chlorella vulgaris	Melt mixing, hot molding	Glycerol, polyvinyl anhydride, maleic anhydride	Microalgal as composites	[28]
Chlorella sp.	Melt mixing, hot molding, compression molding	Polyethylene, maleic anhydride	synthesize new compound	[29]
Nannochloropsis gaditana	Melt mixing, injection molding, twin screw extrusion	Poly (butylene adipate- co terephthalate)	residual biomass, microalgal as composites	[30]
Nannochloropsis sp., *Spirulina* sp., *Scenedesmus* sp.	Melt mixing, compression molding, solvent casting	Corn starch biocomposites	Microalgal as composites	[31]
Spirulina sp., *Chlorella* sp.	Compression molding	Polyethylene polymer	Microalgal as composites	[32]
Spirulina platensis	Compression molding	Gluten	Microalgal as filler	[33]
Spirulina sp.	Compression molding	Poly (butylene succinate)	Microalgal as composites	[34]
Nannochloropsis sp.	Compression molding	Polyethylene	Remove odor problem	[35]
Ankistrodesmus falcatus (NIES-2195), *Chlamydomonas reinhardtii* 11-32A, *Parachlorella kessleri* (NIES-2152), *C. reinhardtii* (DW15), *Scenedesmus obliquus* (NIES-2280), *Chlorella sorokiniana* (NIES-2173), *Chlorella variabilis* (NC-64A), *C. vulgaris* (NIES-227), *Scenedesmus acutus* (NIES-94), *Scenedesmus* sp.	Twin screw extrusion	Glycerol	intracellular starch production microalgae in sulfur-deprived medium	[36]
Chlorella sp.	Solvent casting	Chlorellapolyvinyl alcohol (PVA)	Microalgal as composites	[37]
Spirulina sp.	Solvent casting	Poly (vinyl alcohol) (PVA)	hazardous high-salt microalgal residues	[38]
Microcystis sp.; *Haematococcus pluvialis*	Solvent casting	Polyhydroxybutyrate (PHB)	Using high-rate algal ponds (HRAP) for algal biomass production	[39]

**Based on Onen Cinar, Chong, Kucuker, Wieczorek, Cengiz, and Kuchta [26].*

Table 1.
Development of microalgae-based bioplastics.

intracellular synthesis of biopolymers in microalgae cells, such as polyhydroxybutyrates (PHBs) and starch (**Table 1**) [26].

Although the production of cassava-microalgae-based bioplastics is feasible, no research on cassava bioplastics based on microalgae has been reported. Nevertheless, Cardoso, et al. [40] have developed biobased films from cassava bagasse and *Spirulina platensis.* This biofilm has a total solid content of 7%, with 4% cassava starch, 1% glycerol, and 2% cassava bagasse/*S. platensis*/gelatin mixture. The greatest elongation value was discovered in a mixture of cassava bagasse: *S. platensis*: gelatin (0.34:1.32:0.34). The inclusion of *S. platensis* raised the color (the value of *a*)* and opacity, but the addition of cassava bagasse increased the viscosity. The films' green color makes them perfect for packing meals of the same color.

The study reported that the biobased films made from cassava bagasse and *S. platensis* are applied to Cambuci peppers (*Capcisum* sp.). For 14 days, the peppers are stored at room and refrigerator temperatures. At room temperature, peppers coated with cassava bagasse-*S. platensis* biobased films lose less bulk. The temperature of the refrigerator delays the maturity of the peppers. In addition, cambuci peppers' shelf life can be extended by combining coating with low temperature [41].

4. Cassava wastewater as a substrate for microalgae

The cassava industry generates a large number of byproducts and leftovers that are rich in organic matter and particulate debris [42]. In addition, cassava processing generates a lot of effluents. It should be noted that processing one ton (1000 kg) of cassava roots may result in between 250 and 600 kilos of wastewater [43]. de Carvalho, et al. [44] reported that cassava wastewater has high quantities of numerous mineral nutrients. The ash level of cassava waste ranges between 2.5% and 3.5% [17]. The biological oxygen demand (BOD), chemical oxygen demand (COD), and total suspended solids (TDS) found in cassava wastewater are all above average [45, 46]. Wastewater's excessive TDS, BOD, and COD levels are directly correlated to the organic matter and chemical content of the water, which poses serious threats to the environment and the well-being of living things [47].

Microalgae are dependent on the chemical composition and organic matter of cassava wastewater. COD reflects the quantity of oxygen that can be consumed by reactions in a measured solution. The biological oxygen demand (BOD), also known as biological oxygen need, is the amount of dissolved oxygen required by aerobic biological organisms to decompose organic material present in a given water sample at a particular temperature over a certain time period. BOD and COD can be utilized to assess the effectiveness of wastewater treatment plants [48]. Nitrogen is an important factor in the growth of microalgae. Nitrogen is essential for the formation of DNA, proteins, and pigments [49]. Phosphorus is also an essential component of numerous metabolic activities, such as the transfer of energy across cell membranes and cells, the production of nucleic acids, and the encouragement of cell growth and photosynthetic activity [50].

Microalgae can naturally absorb nutrients from the water in order to grow. Thus, laboratory studies have shown that nutrient removal via microalgae cultivation is possible [51]. Microalgae can be autotrophic or heterotrophic. If they are autotrophic, they obtain their carbon from inorganic molecules. Autotrophs are classified into two types: chemolithotrophic and photoautotrophic (photolithotrophic). Photoautotrophs use light as an energy source, whereas chemoautotrophs (chemolithotrophs) oxidize inorganic substances for energy. Furthermore, heterotrophic microalgae employ organic molecules for growth. Heterotrophs are classified into two groups: photoheterotrophs (photoorganotrophs) and heterotrophs (chemoorganotrophs). Photoheterotrophs use light as an energy source, whereas chemoheterotrophs (chemoorganotrophs) oxidize organic substances for energy. Furthermore, heterotrophic microalgae can absorb complete food particles into food vesicles for processing. They may, on the other hand, be osmotrophic, absorbing dissolved nutrients through the plasma membrane. Some photosynthetic algae are mixotrophic (facultatively heterotrophic). They rely on organic chemicals found in the media [52]. Sorgatto, Soccol, Molina-Aulestia, de Carvalho, de Melo Pereira, and de Carvalho [53] reported that microalgae grew faster in cassava wastewater and produced lipids similar to synthetic mixotrophic cultures. Another research by Nwanko and Agwa (2021) showed that the optimal ratio of cassava peel water to cassava wastewater (CP:CW) for growth was 160:40.

Nutrients in cassava wastewater can be removed using microalgae. For wastewater treatment, the microalgal genera Chlorella, Haemotococcus, Arthrospira, and Dictyosphaerium have been studied [45, 47, 54–57]. Chai, et al. [58] reported that microalgae are effective at removing nitrogen, phosphate, and toxic metals from wastewater. de Faria Ferreira Carraro, Loures, and de Castro [46] demonstrated cyanide removal efficacy approaching 99% and average CO_2 biofixation of 0.19 g L^{-1} from cassava wastewater. The efficacy of wastewater treatment varies according to the species as shown in **Table 2**.

Microalgae-treated wastewater has the potential to be used in pollution removal, agriculture, aquaculture, biogas production, bioproducts, and biomaterials [58, 60]. One method for producing biomass and metabolites at a cheaper cost is to cultivate

Species	Efficiency of waste removal	References
Chlorella minutissima	COD, TS, and nutrient removal efficiency are about 30, 75, and 92%, respectively. In addition, cyanide removal was 99% and average CO_2 biofixation was 0.19 g L^{-1} d^{-1}	[46]
Dictyosphaerium sp.	TS for heterotrophy was 79.32% and for mixotrophy was 89.78%, respectively. For heterotrophy and mixotrophy, the BOD was 72.95% and 89.35%, the COD was 72.19% and 84.03%, and the cyanide level reduces from 450 mg/L to 93.105 (79.31%) and 85.365 mg/L (81.03%), respectively.	[47]
Arthrospira platensis	Over 99% of ammonia, nitrites, and nitrates, could be reduced.	[54]
Chlorella sorokiniana WB1DG and *C. sorokiniana* P21	COD, total phosphorous (TP), and total inorganic nitrogen (TIN) are reduced by 63.42, 91.68, and 70.66%, respectively, in P21 culture. While WB1DG culture reduced COD, TP, and TIN by up to 73.78, 92.11, and 67.33%, respectively.	[59]
C. sorokiniana WCF	COD and nutrient removal efficiency in WCF supernatant ranged between 80 and 94%.	[45]
Haematococcus pluvialis and *Neochloris oleoabundans*	COD, TN, and TP are reduced by 60.80, 51.06, and 54.68%, respectively, in HP culture. While in NO culture reduced COD, TN, and TP by up to 69.16, 58.19, and 69.84, respectively.	[53]

Table 2.
Microalgae waste removal efficiency.

Type	Species	Metabolites	References
Cassava water processing	*Neochloris oleoabundans* UTEX 1185 and *Haematococcus pluvialis* SAG 34 – 1b and	Fatty acid methyl esters (FAME)	[53]
Non-detoxified cassava bagasse hydrolysate (CBH)	*Chlorella pyrenoidosa* and yeast *Rhodotorula glutinis*	FAME, sugar	[62]
Cassava water waste	*H. pluvialis*	Astaxanthin	[56]
Cassava water waste	*Dictyosphaerium* sp. LC172264	Lipid	[47]

Table 3.
Metabolites formed during the processing of cassava waste.

microalgae in wastewater [53, 61]. **Table 3** depicts the cassava wastewater treatment product based on microalgae. Wastewater from cassava processing can be utilized to cultivate microalgae, which can be used to produce biodiesel or biogas [44, 63]. Microalgae can also be used as a bio stabilizer to boost biogas production from cassava starch effluent [64]. The use of cassava wastewater can also increase astaxanthin accumulation and reduce the toxicity caused by this agro-industrial effluent in *H. pluvialis* [56].

5. Cationic cassava starch as flocculants for microalgae harvesting

The most difficult challenge in collecting biomass for subsequent usage is efficiently collecting microalgae biomass from their growth substrate. According to Al-Hattab, et al. [65], it accounted for 20–30% of overall biomass production expenses. We are aware that harvesting using filtration techniques is one of the most cost-effective methods available, despite the fact that it is limited to microalgae with a large size, such as *Arthrospira* sp. This method is widely used, even in industrial production. Indeed, microalgae harvesting strategies consider species characteristics, size, density, downstream biomass processing, and sometimes medium recycling needs [66–68]. The greater the size or length of the microalgae, the greater the number of potential screening options available. In actuality, single-cell and small-sized microalgae predominate in our environment, and we constantly interact with them. These factors encourage the development of harvesting systems that are sustainable, cost-effective, suited for industrial production, and safe based on the final product's intended use.

Researchers have studied various methods of harvesting microalgae biomass, including filtration with specific treatments, centrifugation, dispersed air flotation, sedimentation, flocculation, bioflocculation, coagulation, and fluidic oscillation [67, 69–71]. The previously researched bioflocculation harvesting method was regarded as a viable approach for gathering microalgal biomass since it was acceptable with regard to sustainability, large-scale production, low-cost energy, and environmental friendliness [47]. When bioflocculant is utilized, however, microbially contaminated biomass products may be produced [72]. In contrast to microbial-based flocculants, plant-based coagulants enable the harvesting of microalgal biomass that is biodegradable and less contaminated by microorganisms, as mentioned previously. Apart from *Moringa oleifera*, neem, cactus, orange (peels), pomegranate (peels), banana (peels), *Canavalia ensiformis*, *Strychnos potatorum*, and *Azadirachta indica*, a cassava starch-based coagulant has previously been examined for microalgal harvesting [73–77]. Moreover, considering biodegradability, renewability, and cost-effectiveness, it becomes attractive to use cassava starch as a natural flocculant [78].

The flocculation harvesting concept is driven by the fact that most microalgae cells have a negative surface wall charge; hence, their stability in suspension is a result of their mutual repulsion [74]. With a positive charge, flocculant The so-called cationic flocculant can absorb and react with the naturally negatively charged microalgae, which results in charge cancelation and cell agglomeration, also known as charge neutralization [79, 80]. Natural flocculants that are positively charged are presumably able to agglomerate the microalgae cells in suspension by means of the mechanism [81]. Moreover, another flocculation mechanism, namely bridging, occurs when natural flocculants with long polymers and a large molecular weight are unable to build polymer bridges with negatively charged ions [82]. Charged cationic biopolymers are also using the mechanism to construct the bridge between the cell walls by means of neutralization of the charge or through electrostatic path agglomeration [83]. Bioflocculant mechanisms are illustrated in **Figure 4**.

Cassava starch has been reported as a flocculant in numerous applications, including kaolin suspension, water treatment, water purification, and wastewater treatment [84–88]. X-ray diffraction (XRD), Fourier-transform infrared spectroscopy (FTIR), and scanning electron microscopy (SEM) characterizations and chemical analysis confirmed that double helix configurations in cassava starch result from the amorphous crystalline distribution of starches and influence the growth of flocs and, thus, sedimentation [89, 90]. Cassava starch with a range of unique modified structures has been used to harvest valuable microalgae from their culture medium due to its promising application. When applied to *Chlorella* sp., the harvesting efficiency of cationic cassava starch in composite magnetic form was able to reach 98.09% with 1.67 g microalgae per g flocculant harvesting capacity [78]. In addition, according to Chittapun, Jangyubol, Charoenrat, Piyapittayanun, and Kasemwong [13], a 1000 mg/L dose of

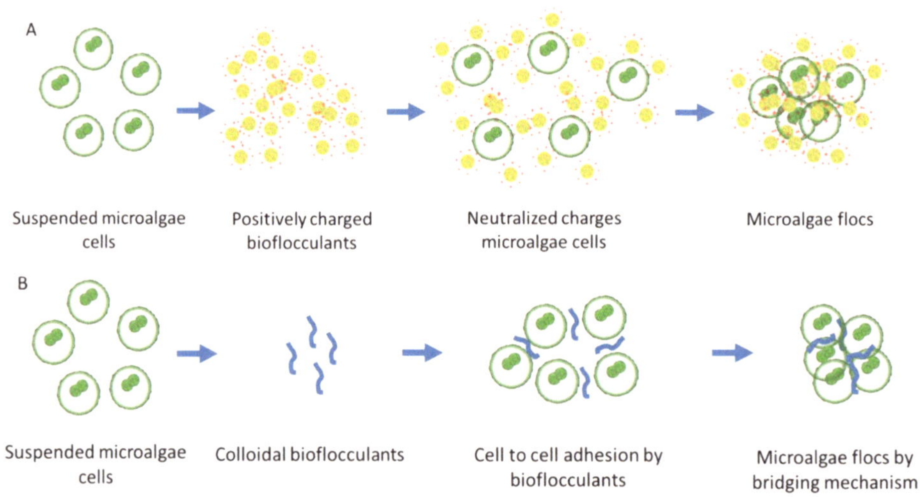

Figure 4.
Mechanisms of bioflocculants that interact with microalgae cells: A. charge neutralization mechanism; B. bridging mechanism (reconstructed according to Ogbonna and Nwoba [47]).

Flocculant	Microalgae species	Operating condition	Dose; Time (min.)	Harvesting efficiency/ Harvesting capacity (g algae/g flocculant)	Cell density	Marine (M)/ Freshwater (F)	Flocculant form	Microalgae properties after recovery	Ref.
Cassava starch	*Chlorella* sp. (TISTR8236)	pH 9.5; 0.76 DS	500 mg/L; 2	98.09%/1.67 ± 0.01	1 g/L	F	Composites, magnetic-cationic cassava starch	NA	[78]
Cassava starch	*Chlorella* sp. (TISTR8236)	pH 10; 0.76 DS	200 mg/L, 2	93.77 ± 0.26 (% ± SD)/4.69 ± 0.01	1 g/L	F	Composites, magnetic-cationic cassava starch	NA	[78]
Cassava starch	*Chlorella* sp. (TISTR8236)	pH 9, 0.04 DS,	1000 mg/L; 2	92.86 (% ± SD)/1.58 ± 0.01	1 g/L	F	Commercial cationic starch	NA	[13]
Cassava starch	*Chlorella* sp. (TISTR8236)	pH 9; 0.040 DS	500 mg/L; 2	67.89 ± 0.48 (% ± SD) / 2.31 ± 0.02	1 g/L	F	Commercial cationic starch	NA	[13]
Cassava starch	*Chlorella* sp. (TISTR8236)	pH 9.5; 0.040 DS	200 mg/L; 2	60.12 ± 0.74 (% ± SD)/2.56 ± 0.03	1 g/L	F	Commercial cationic starch	NA	
Cassava starch	*Chlorella* sp. (TISTR8236)	pH 9.5; 0.04 DS	200 mg/L; 2	56.65 ± 0.30 (% ± SD)/2.41 ± 0.01	1 g/L	F	Composites, magnetic-cationic cassava starch	NA	[13]
Cassava starch	*Chlorella* sp. (TISTR8236)	pH 9.5; 0.76 DS	200 mg/L; 2	80.15 ± 0.29 (% ± SD)/3.41 ± 0.01	1 g/L	F	Composites, magnetic-cationic cassava starch	NA	[13]
Potato starch	*Scenedesmus dimorphus.*	pH 7; 0.14 and 0.31 DS	10 mg/L; 90	> 95% /	0.12 g/L	F	Cationic starch batches were synthesized using CHPTAC	NA	[91]
Potato starch	*S. dimorphus.*	pH 7; 0.41 and 0.64 DS	10 mg/L; 90	>95%	0.12 g/L	F	Cationic starch was created using EPTAC	NA	[91]
Potato starch	*Scenedesmus obliquus*	pH 7; 0.82 DS	0.0053 coagulant/algae ratio; 72	85% /	0.2–0.25 g/L	F	Cationic starch was using MAPTAC	NA	[92]

Flocculant	Microalgae species	Operating condition	Dose; Time (min.)	Harvesting efficiency/ Harvesting capacity (g algae/g flocculant)	Cell density	Marine (M)/ Freshwater (F)	Flocculant form	Microalgae properties after recovery	Ref.
Potato starch	*S. obliquus*	pH 7; 0.82 DS	1.4 coagulant/ algae ratio; 72	60% /	0.035–0.05 g/L	F	Cationic starch was using MAPTAC	NA	[92]

**DS=Degree of Substitutions.*

Table 4.
Characteristics of cassava and other bioflocculants for algae harvesting.

Ref.	Flocculant	Cost	Level	Toxicity	Application	Advantages	Drawbacks
[78]	Composites, magnetic-cationic cassava starch	NA	Laboratory	Potentially contaminated	Another source of raw materials for making biofuels.		Low detachment ability may limit the use of collected algal cells.
[13]	Commercial cationic starch	NA	Laboratory	Potentially contaminated			
[13]	Commercial cationic starch	NA	Laboratory	Not	It is more affordable to buy cationic cassava starch from the market. In contrast to the composite, the final biomass was not contaminated.		
[13]	Composites, magnetic-cationic cassava starch	NA	Laboratory	Potentially contaminated			
[91]	Cationic potato starch batches were synthesized using CHPTAC	NA	Laboratory	NA		Low dosage	
[92]	Cationic potato starch was synthesized using epoxide equivalent 2,3-epoxypropyl trimethyl ammonium chloride (EPTAC)	NA	Laboratory	NA		Low dosage	
[92]	Cationic potato starch	NA	Laboratory	NA			
[92]	Cationic potato starch	NA	Wastewater from Lagoons	NA			

Table 5.
Application of cassava flocculants for microalgae harvesting.

commercial cationic starch enabled harvesting of the same species of microalgae at the maximum harvesting capacity of 92.86% (**Table 4**). For comparison, potato starch, on the other hand, has been applied for *Scenedesmus dimorphus* flocculation. Cationic starch was used to precipitate more than 95% of the microalgae. Cationic starch was generated using N-(3-chloro-2-hydroxypropyl) trimethyl ammonium chloride (CHPTAC), and epoxide equivalent 2,3-epoxypropyl trimethyl ammonium chloride (EPTC) was used to make cationic starch [91]. While Anthony and Sims [92] investigated different forms of potato starch flocculant using 3-methacryloyl aminopropyl trimethyl ammonium chloride (MAPTAC) and discovered disparate results as shown in **Table 4**. This research has already demonstrated the effects of dose, pH, and degree of substitution on the efficiency of cassava starch as a harvesting aid for microalgae. Additionally, the ionic strength of the medium, hydrophobicity and net charge, phase of cellular growth, CO_2 concentration, particle size, and zeta potential will influence flocculation processes [93, 94].

The use of cassava starch as a microalga harvesting agent has already been tested in the laboratory as shown in **Table 5**. The volume of microalgae cultures that have been employed in numerous industries (food, feed, and cosmetics). According to FAO [95], 56,456 tons of microalgae were produced, with 99.56% coming from *Spirulina* (Arthrospira) and the remaining 248 tons derived from other single-celled microalgae such as *Haematococcus pluvialis, Chlorella vulgaris*, *Tetraselmis* spp., and *Dunaliella salina*. One of the reasons for the low production of other single-celled macroalgae could be the inefficiency of the harvesting technique. In light of this, it appears that studies on flocculation harvesting techniques, including the use of cassava starch as a natural material-based flocculant agent, are still in great demand. However, laboratory research on cassava starch as a harvesting agent for microalgae is still limited to microalgae and cassava species. Only two literatures, namely Jangyubol, Kasemwong, Charoenrat and Chittapun [78] and Chittapun, Jangyubol, Charoenrat, Piyapittayanun, and Kasemwong [13] have intensively investigated cassava starch as a flocculant agent. Some additional research is required, such as increasing microalgae volume levels, toxicity analysis, marine microalgae species utilization, end product nutrient analysis, and until microalgae production is economically feasible. Cassava starch's economic worth, renewable availability, and biodegradability have attracted industrial levels despite the fact that starch is only the planet's second most plentiful natural polymer [13].

6. Future prospective

The integration of the cassava industry and the microalgae biorefinery is a promising and sustainable technology. Using cassava waste to grow microalgae can help to reduce the environmental impact of the cassava flour industry. Additionally, using cassava waste as a substrate for microalgae can reduce manufacturing costs. Microalgae can generate useful products comprising protein, lipids, antioxidants, and vitamins. In addition, harvesting microalgae with cassava starch is highly effective and minimizes the cost of producing microalgae biomass. In addition, the development of fortified staple foods is necessary for the future improvement of public health. Trends in health and fitness increase demand for microalgae and cassava products. Microalgae can be added to cassava products to improve their nutritional value. The addition of microalgae to cassava products can help the sustenance of the community. Furthermore, the addition of microalga to baby food items helps prevent infant stunting. Moreover, biomass derived from microalgae and cassava can be used to produce plant-based proteins. The development of high-protein and mineral-rich plant-based proteins is primarily motivated by the desires of vegetarian and vegan societies.

In conclusion, the integration of the cassava industry with microalgae biorefinery has numerous environmental and social benefits. It contributes to the implementation of clean technology within the cassava industry. Additionally, it adds to the nourishment of the population. Responsible consumption and production, ending hunger, protecting terrestrial and marine ecosystems, and preserving biodiversity are just a few of the Sustainable Development Goals that benefit from this work.

Author details

Ardiba Rakhmi Sefrienda, Dedy Kurnianto, Jasmadi Jasmadi and Andri Frediansyah*
Research Center for Food Technology and Processing (PRTPP), National Research and Innovation Agency (BRIN), DI. Yogyakarta, Indonesia

*Address all correspondence to: andr042@brin.go.id

References

[1] Parmar A, Sturm B, Hensel O. Crops that feed the world: Production and improvement of cassava for food, feed, and industrial uses. Food Security. 2017;**9**(5):907-927

[2] Frediansyah A. Microbial Fermentation as Means of Improving Cassava Production in Indonesia. 2018

[3] Li S, Cui Y, Zhou Y, Luo Z, Liu J, Zhao M. The industrial applications of cassava: Current status, opportunities and prospects. Electronic. The Journal of the Science of Food and Agriculture. London: SCI by John Wiley & Sons Ltd. 2017:1097-1110

[4] Savitri S. Cassava-based packaging for a better future. Available from: https://www.industrysourcing.com/article/Cassava-based-packaging-for-a-better-future (27 November)

[5] Tsige TZ, Basa B, Herago T. Medicinal, nutritional and anti-nutritional properties of cassava (Manihot esculenta): A review. 2019:34-46

[6] Shigaki T. In: Caballero B, Finglas PM, Toldrá FB T-Eo F a H, editors. Cassava: The Nature and Uses. Oxford: Academic Press; 2016. pp. 687-693

[7] Şahin OI. Effect of spirulina biomass fortification for biscuits and chocolates. In: Turkish Journal of Agriculture - Food Science and Technology. 2019

[8] Abd El-Salam AM, Morsy OM, Abd El Mawla EM. Production and evaluation crackers and instant noodles supplement with spirulina algae. Current Science International. Pakistan. 2017;**6**:908-919

[9] Barakat EH, El-Kewaisny NM, Salama AA. Chemical and nutritional evaluation of fortified biscuits with dried spirulina algae. Journal of Food and Diary Sciences. 2016;**7**:167-177

[10] Freitas BCB, Santos TD, Moreira JB, Zanfonato K, Morais MG, Costa JAV. Novel foods: A meal replacement shake and a high-calorie food supplemented with spirulina biomass. International Food Research Journal. 2019;**26**:59-65

[11] Navacchi M, Carvalho JC, Takeuchi K, Danesi E. Development of cassava cake enriched with its own bran and Spirulina platensis. Acta Scientiarum Technology. 2012;**34**:465-472

[12] Rabelo S, Lemes A, Takeuchi K, Frata M, Carvalho JC, Danesi E. Development of cassava doughnuts enriched with Spirulina platensis biomass. Brazilian Journal of Food Technology. 2013;**16**:42-51

[13] Chittapun S, Jangyubol K, Charoenrat T, Piyapittayanun C, Kasemwong K. Cationic cassava starch and its composite as flocculants for microalgal biomass separation. International Journal of Biological Macromolecules. 2020;**161**:917-926

[14] Achi CG, Coker AO, Sridhar MKC. In: Ghosh SK, editor. Cassava Processing Wastes: Options and Potentials for Resource Recovery in Nigeria, Utilization and Management of Bioresources, Singapore. Singapore: Springer Singapore; 2018. pp. 77-89

[15] Olukanni DO, Olatunji TO. Cassava waste management and biogas generation potential in selected local government areas in Ogun state, Nigeria. Recycling. 2018;**3**(4):58

[16] Zhang L, Chen L, Wang J, Chen Y, Gao X, Zhang Z, et al. Attached

cultivation for improving the biomass productivity of Spirulina platensis. Electronic. 2015:1873-2976

[17] Veiga JPS, Valle TL, Feltran JC, Bizzo WA. Characterization and productivity of cassava waste and its use as an energy source. Renewable Energy. 2016;**93**:691-699

[18] Kusumaningrum A, Solikhah M, Herawati ERN, Susanto A, Ariani D. Gluten-free snacks cheese stick based on mocaf (modified cassava) flour: Properties and consumer acceptance. IOP Conference Series: Earth and Environmental Science. 2019;**251**(1)

[19] Putri DP, Yulianti LE, Afifah N. Accelerated shelf life testing of mocatilla chip using critical moisture content approach and models of sorption isotherms. IOP Conference Series: Materials Science and Engineering. 2021;**1011**:1-8

[20] Sefrienda AR, Ariani D, Fathoni A. Karakteristik Mi berbasis tepung ubi kayu termodifikasi (MOCAF) yang Diperkaya Ekstrak Wortel (Daucus Carota). Jurnal Riset Teknologi Industri. 2020;**14**(2):133-141

[21] Setiaboma W, Kristanti D, Afifah N. Pendugaan Umur Simpan Kukis Mocaf dengan metode akselerasi berdasarkan kadar air kritis. Jurnal Riset Teknologi Industri. 2020;**14**(2):167-167

[22] Ariani RP, Ekayani IAPH, Masdarini L. Processing MOCAF into pie susu with the addition of super food 'spirulina'. Journal of Physics: Conference Series. 2021;**1810**(1):012078

[23] Al-Baarri A, Widayat W, Aulia R, Prahasiwi E, Mawarid A, Pangestika W, et al. The hardness analysis of noodles made from modified cassava flour, spirulina (Spirulina platensis) and basil leaves extract (Ocimum sanctum L.). IOP Conference Series: Earth and Environmental Science. 2021;**653**:12051

[24] Verma R, Chauhan N, Singh B, Chandra S, Author C, Sengar R. Cassava processing and its food application: A review. The Pharma Innovation Journal. 2022;**SP-11**(5):415-422

[25] Ashter SA. Introduction. In: Ashter SA, editor. Introduction to Bioplastics Engineering. Oxford: William Andrew Publishing; 2016. pp. 1-17

[26] Onen Cinar S, Chong ZK, Kucuker MA, Wieczorek N, Cengiz U, Kuchta K. Bioplastic production from microalgae: A review. Electronic. 2020:1660-4601

[27] Ismail D, Gozan M, Noviasari C. The effect of glycerol addition as plasticizer in Spirulina platensis based bioplastic. E3S Web of Conferences. 2018;**67**:03048

[28] Ismail D, Khalis SA. The effect of compatibilizer addition on Chlorella vulgaris microalgae utilization as a mixture for bioplastic. E3S Web of Conferences. 2018;**67**:03047

[29] Otsuki T, Zhang F, Kabeya H, Hirotsu T. Synthesis and tensile properties of a novel composite of chlorella and polyethylene. Journal of Applied Polymer Science. 2004;**92**:812-816

[30] Torres S, Navia R, Campbell Murdy R, Cooke P, Misra M, Mohanty AK. Green composites from residual microalgae biomass and poly(butylene adipate-co-terephthalate): Processing and plasticization. ACS Sustainable Chemistry & Engineering. 2015;**3**(4):614-624

[31] Fabra MJ, Martínez-Sanz M, Gómez-Mascaraque LG, Gavara R, López-Rubio A. Structural and physicochemical

characterization of thermoplastic corn starch films containing microalgae. Electronic. 2018:1879-1344

[32] Zeller MA, Hunt R, Jones A, Sharma S. Bioplastics and their thermoplastic blends from spirulina and chlorella microalgae. Journal of Applied Polymer Science. 2013;**130**(5):3263-3275

[33] Ciapponi RA-OX, Turri S, Levi M. Mechanical reinforcement by microalgal biofiller in novel thermoplastic biocompounds from plasticized Gluten. Materials. 2019:1476. DOI: 10.3390/ma12091476

[34] Zhu N, Ye M, Shi D, Chen M. Reactive compatibilization of biodegradable poly(butylene succinate)/spirulina microalgae composites. Macromolecular Research. 2017;**25**(2):165-171

[35] Wang K, Mandal A, Ayton E, Hunt R, Zeller MA, Sharma S. Chapter 6 - Modification of protein rich algal-biomass to form bioplastics and odor removal. In: Singh Dhillon G, editor. Protein Byproducts. Canada: Academic Press; 2016. pp. 107-117

[36] Mathiot C, Ponge P, Gallard B, Sassi JF, Delrue F, Le Moigne N. Microalgae starch-based bioplastics: Screening of ten strains and plasticization of unfractionated microalgae by extrusion. Electronic. 2019:1879-1344

[37] Sabathini HA, Windiani L, Ismail D, Gozan M. Mechanical physical properties of chlorella-PVA based bioplastic with ultrasonic homogenizer. E3S Web of Conferences. 2018;**67**:03046

[38] Zhang C, Wang C, Cao G, Wang D, Ho SH. A sustainable solution to plastics pollution: An eco-friendly bioplastic film production from high-salt contained Spirulina sp. residues. Electronic. 2020:1873-3336

[39] Abdo SM, Ali GH. Analysis of polyhydroxybutrate and bioplastic production from microalgae. Bulletin of the National Research Centre. 2019;**43**(1):97

[40] Cardoso T, Esmerino LA, Bolanho BC, Demiate IM, Danesi EDG. Technological viability of biobased films formulated with cassava by-product and Spirulina platensis. In: Journal of Food Process Engineering. Vol. 42. USA: John Wiley & Sons, Ltd; 2019. p. e13136

[41] Cardoso T, Demiate I, Danesi E. Biodegradable films with Spirulina platensis as coating for Cambuci peppers (capsicum sp.). American Journal of Food Technology. 2017;**12**:236-244

[42] Zhang M, Xie L, Yin Z, Khanal SK, Zhou Q. Biorefinery approach for cassava-based industrial wastes: Current status and opportunities. Bioresource Technology. 2016;**215**:50-62

[43] Oghenejoboh KM, Orugba HO, Oghenejoboh UM, Agarry SE. Value added cassava waste management and environmental sustainability in Nigeria: A review. Environmental Challenges. 2021;**4**:100127

[44] de Carvalho JC, Borghetti IA, Cartas LC, Woiciechowski AL, Soccol VT, Soccol CR. Biorefinery integration of microalgae production into cassava processing industry: Potential and perspectives. Bioresource Technology. 2018;**247**:1165-1172

[45] Melo JM, Telles TS, Ribeiro MR, de Carvalho Junior O, Andrade DS. Chlorella sorokiniana as bioremediator of wastewater: Nutrient removal, biomass production, and potential profit. Bioresource Technology Reports. 2022;**17**:100933

[46] de Faria Ferreira Carraro C, Loures CCA, de Castro JA. Microalgae technique for bioremediation treatment of cassava wastewater. Water, Air, & Soil Pollution. 2021;**232**(7):281

[47] Ogbonna CN, Nwoba EG. Bio-based flocculants for sustainable harvesting of microalgae for biofuel production. A review. Renewable and Sustainable Energy Reviews. 2021;**139**:110690

[48] Li K, Liu Q, Fang F, Luo R, Lu Q, Zhou W, et al. Microalgae-based wastewater treatment for nutrients recovery: A review. Bioresource Technology. 2019;**291**:121934

[49] Markou G, Angelidaki I, Georgakakis D. Microalgal carbohydrates: An overview of the factors influencing carbohydrates production, and of main bioconversion technologies for production of biofuels. Applied Microbiology and Biotechnology. 2012;**96**(3):631-645

[50] Liyanaarachchi VC, Nishshanka GKSH, Premaratne RGMM, Ariyadasa TU, Nimarshana PHV, Malik A. Astaxanthin accumulation in the green microalga Haematococcus pluvialis: Effect of initial phosphate concentration and stepwise/continuous light stress. Biotechnology Reports. 2020;**28**:e00538

[51] Nguyen TDP, Le TVA, Show PL, Nguyen TT, Tran MH, Tran TNT, et al. Bioflocculation formation of microalgae-bacteria in enhancing microalgae harvesting and nutrient removal from wastewater effluent. Bioresource Technology. 2019;**272**:34-39

[52] Sluiman H. Phycology. In: Lee RE, editor. Edinburgh Journal of Botany. 4th ed. United Kingdom: Trustees of the Royal Botanic Garden Edinburgh; 2009. p. 66

[53] Sorgatto VG, Soccol CR, Molina-Aulestia DT, de Carvalho MA, de Melo Pereira GV, de Carvalho JC. Mixotrophic cultivation of microalgae in cassava processing wastewater for simultaneous treatment and Production of lipid-rich biomass. In Fuels. 2021;**2**:521-532

[54] Araujo GS, Santiago CS, Moreira RT, Dantas Neto MP, Fernandes FAN. Nutrient removal by Arthrospira platensis cyanobacteria in cassava processing wastewater. Journal of Water Process Engineering. 2021;**40**:101826

[55] Liu J, Zheng QJ, Ma QX, Gadidasu KK, Zhang P. Cassava genetic transformation and its application in breeding. Journal of Integrative Plant Biology. 2011;**53**:552-569

[56] Coutinho Rodrigues OH, Itokazu AG, Rörig L, Maraschin M, Corrêa RG, Pimentel-Almeida W, et al. Evaluation of astaxanthin biosynthesis by Haematococcus pluvialis grown in culture medium added of cassava wastewater. International Biodeterioration & Biodegradation. 2021;**163**:105269

[57] Padri M, Boontian N, Teaumroong N, Piromyou P, Piasai C. Co-culture of microalga Chlorella sorokiniana with syntrophic streptomyces thermocarboxydus in cassava wastewater for wastewater treatment and biodiesel production. Bioresource Technology. 2022;**347**:126732

[58] Chai WS, Tan WG, Halimatul Munawaroh HS, Gupta VK, Ho S-H, Show PL. Multifaceted roles of microalgae in the application of wastewater biotreatment: A review. Environmental Pollution. 2021;**269**:116236

[59] Padri M, Boontian N, Teaumroong N, Piromyou P, Piasai C. Application of two indigenous strains of microalgal

Chlorella sorokiniana in cassava biogas effluent focusing on growth rate, removal kinetics, and harvestability. Water. 2021;**13**:1-21

[60] Al-Jabri H, Das P, Khan S, Thaher M, AbdulQuadir M. Treatment of wastewaters by microalgae and the potential applications of the produced biomass—A review. Water. 2021;**13**:1-26

[61] Ren Z, Jia B, Zhang G, Fu X, Wang Z, Wang P, et al. Study on adsorption of ammonia nitrogen by iron-loaded activated carbon from low temperature wastewater. Chemosphere. 2021;**262**:127895

[62] Liu L, Chen J, Lim P-E, Wei D. Enhanced single cell oil production by mixed culture of Chlorella pyrenoidosa and Rhodotorula glutinis using cassava bagasse hydrolysate as carbon source. Bioresource Technology. 2018;**255**:140-148

[63] Neves C, Maroneze MM, dos Santos AM, Francisco ÉC, Wagner R, Zepka LQ, et al. Cassava processing wastewater as a platform for third generation biodiesel production. In Scientia Agricola. 2016;**73**:412-416

[64] Budiyono B, Kusworo T. Microalgae for stabilizing biogas Production from cassava starch wastewater. International Journal of Waste Resources (IJWR). 2012;**2**

[65] Al-Hattab M, Ghaly A, Hammouda A. Microalgae harvesting methods for industrial production of biodiesel: Critical review and comparative analysis. Journal of Fundamentals of Renewable Energy and Applications. 2015;**5**:1-26

[66] Uduman N, Qi Y, Danquah MK, Forde GM, Hoadley AFA. Dewatering of microalgal cultures: A major bottleneck to algae-based fuels. Journal of Renewable and Sustainable Energy. 2010;**2**:1

[67] Rawat I, Ranjith Kumar R, Mutanda T, Bux F. Dual role of microalgae: Phycoremediation of domestic wastewater and biomass production for sustainable biofuels production. Applied Energy. 2011;**88**(10):3411-3424

[68] Amaro HM, Guedes AC, Malcata FX. Advances and perspectives in using microalgae to produce biodiesel. Applied Energy. 2011;**88**(10):3402-3410

[69] Kim L, Kwon K, Oh Y. Effects of harvesting method and growth stage on the flocculation of the green alga Botryococcus braunii. Letters in Applied Microbiology. 1998;**27**(1):14-18

[70] Pittman JK, Dean AP, Osundeko O. The potential of sustainable algal biofuel production using wastewater resources. Bioresource Technology. 2011;**102**(1):17-25

[71] Letelier-Gordo CO, Holdt SL, De Francisci D, Karakashev DB, Angelidaki I. Effective harvesting of the microalgae i via bioflocculation with cationic starch. Bioresource Technology. 2014;**167**:214-218

[72] Barros AI, Gonçalves AL, Simões M, Pires JCM. Harvesting techniques applied to microalgae: A review. Renewable and Sustainable Energy Reviews. 2015;**41**:1489-1500

[73] Teixeira C, Teixeira P. Evaluation of the flocculation efficiency of Chlorella vulgaris mediated by Moringa oleifera seed under different forms: Flour, seed cake and extracts of flour and cake. Brazilian Journal of Chemical Engineering. 2017;**34**:1

[74] Behera B, Balasubramanian P. Natural plant extracts as an economical

and ecofriendly alternative for harvesting microalgae. Bioresource Technology. 2019;**283**:45-52

[75] Díaz-Santos E, Vila M, de la Vega M, León R, Vigara J. Study of bioflocculation induced by Saccharomyces bayanus var. uvarum and flocculating protein factors in microalgae. Algal Research. 2015;**8**:23-29

[76] Abdul Razack S, Duraiarasan S, Santhalin Shellomith AS, Muralikrishnan K. Statistical optimization of harvesting Chlorella vulgaris using a novel bio-source, Strychnos potatorum. Biotechnology Reports. 2015;**7**:150-156

[77] Ali M, Mustafa A, Saleem M. Comparative study between indigenous natural coagulants and alum for microalgae harvesting. Arabian Journal for Science and Engineering. 2019;**44**(7):6453-6463

[78] Jangyubol K, Kasemwong K, Charoenrat T, Chittapun S. Magnetic–cationic cassava starch composite for harvesting chlorella sp. TISTR8236. Algal Research. 2018;**35**:561-568

[79] Salim S, Bosma R, Vermuë MH, Wijffels RH. Harvesting of microalgae by bio-flocculation. Journal of Applied Phycology. 2011;**23**(5):849-855

[80] Salim S, Kosterink NR, Tchetkoua Wacka ND, Vermuë MH, Wijffels RH. Mechanism behind autoflocculation of unicellular green microalgae Ettlia texensis. Journal of Biotechnology. 2014;**174**:34-38

[81] Aljuboori AHR, Idris A, Abdullah N, Mohamad R. Production and characterization of a bioflocculant produced by aspergillus flavus. Bioresource Technology. 2013;**127**:489-493

[82] Aljuboori AHR, Uemura Y, Osman NB, Yusup S. Production of a bioflocculant from aspergillus Niger using palm oil mill effluent as carbon source. Bioresource Technology. 2014;**171**:66-70

[83] Vandamme D, Foubert I, Muylaert K. Flocculation as a low-cost method for harvesting microalgae for bulk biomass production. Trends in Biotechnology. 2013;**31**(4):233-239

[84] Razali MAA, Ariffin A. Polymeric flocculant based on cassava starch grafted polydiallyldimethylammonium chloride: Flocculation behavior and mechanism. Applied Surface Science. 2015;**351**:89-94

[85] Villabona-Ortíz A, Tejada-Tovar C, Millán-Aníbal M, Granados-Conde C, Ortega-Toro R. Reduction of turbidity in waters using cassava starch as a natural coagulant. PRO. 2021;**19**:1

[86] Lugo-Arias J, Lugo-Arias E, Ovallos-Gazabon D, Arango J, de la Puente M, Silva J. Effectiveness of the mixture of nopal and cassava starch as clarifying substances in water purification: A case study in Colombia. Heliyon. 2020;**6**(6):e04296

[87] Jiraprasertkul W, Nuisin R, Jinsart W, Kiatkamjornwong S. Synthesis and characterization of cassava starch graft poly(acrylic acid) and poly[(acrylic acid)-co-acrylamide] and polymer flocculants for wastewater treatment. Journal of Applied Polymer Science. 2006;**102**:2915-2928

[88] Meshram MW, Patil VV, Mhaske ST, Thorat BN. Graft copolymers of starch and its application in textiles. Carbohydrate Polymers. 2009;**75**(1):71-78

[89] Chávez-Salazar A, Bello-Pérez LA, Agama-AcevedoE, Castellanos-GaleanoFJ, Álvarez-Barreto CI, Pacheco-Vargas G.

Isolation and partial characterization of starch from banana cultivars grown in Colombia. International Journal of Biological Macromolecules. 2017;**98**:240-246

[90] Ashri A, Yusof MSM, Jamil MS, Abdullah A, Yusoff SFM, Arip MNM, et al. Physicochemical characterization of starch extracted from Malaysian wild yam (Dioscorea hispida Dennst.). Emirates Journal of Food and Agriculture. 2014;**26**(8)

[91] Hansel PA, Guy Riefler R, Stuart BJ. Efficient flocculation of microalgae for biomass production using cationic starch. Algal Research. 2014;**5**:133-139

[92] Anthony R, Sims R. Cationic starch for microalgae and total phosphorus removal from wastewater. Journal of Applied Polymer Science. 2013;**130**(4):2572-2578

[93] Papazi A, Makridis P, Divanach P. Harvesting Chlorella minutissima using cell coagulants. Journal of Applied Phycology. 2010;**22**(3):349-355

[94] Shelef G, Sukenik A, Green M. Microalgae Harvesting and Processing: A Literature Review. Haifa, Israil: Technion Research and Development Foundation ltd.; 1984

[95] FAO Fishery and Aquaculture Statistics. Available from: www.fao.org/fishery/statistics/software/fishstatj/en

Chapter 9

Use of Cassava in Chicken Diet

Tagesse Tadesse

Abstract

Chicken production has negatively been affected by continuous increment in conventional energy-source feed ingredients due to the competition between human beings and animals globally. Cereal crops, their byproducts, and leftovers from households are among the frequently accessible sources of chicken feed. Poultry industry has been affected by a shortage and increasing cost of conventional feed resources. Various non-conventional feed resources have been reported to solve this problem. Tuber and root crops are among the alternative feed resources and can be substituted at varying quantities in chickens' diets. Among the root and tuber crops that can be included in the diet of chickens is cassava. The tuber of cassava can be cut into chunks, dried, milled, or pelletized and added to the diet of chickens. It can also be added to the diet of broilers, and it can substitute 50% of the maize in their diet without negatively impacting their performance. Adding 40% cassava flour or 20% cassava peel meal to the diet of layer chickens is also beneficial for their ability to lay eggs. Generally, different parts of cassava can be included at different amounts in the diets of chickens.

Keywords: cassava, chicken, feed intake, non-conventional feed, substitution

1. Introduction

African chicken production survives by scavenging and, other than the occasional feeding of household garbage to the chickens and under other circumstances, with the addition of grain to the feed. Due to low input levels and numerous issues with village chicken production, the entire standard of chicken production in developing nations like Ethiopia is primarily of the scavenging type and is typically inefficient [1]. Chickens are a simple way to generate family income and opportunities for job creation with relatively low resource investment and readily available family labor because of their tendency to adapt to most regions of the world, their rapid growth rate, their quick breeding rate compared to most other livestock, or the short generation span. When compared to other domestic animals, chicken is an incredible resource for chicken farming and for household usage as a protein-rich source of animal food for human consumption [2]. The production of chicken is challenged by various factors. The major problem that affects the poultry industries in the tropics is the increasing cost of feed ingredients, such as maize and soya bean meal. The seasonal instability in the supply of conventional feed ingredients requires alternative energy sources to be explored to ensure optimum performance of the chickens [3].

The main challenges for livestock production in the tropics and sub-tropical countries are inadequate feed sources and low quality of existing feed resources [4]. Due to insufficient production of conventional feedstuffs for livestock feeding, the majority of developing nations struggles to provide their animals with enough to feed. Both humans and animals compete over the few amounts of concentrated feedstuffs they produce each year. As a result, the production of livestock in these nations frequently faces a significant problem due to the lack of feed resources [5]. The use of non-conventional feedstuffs in animal feeding can help to solve this problem. In this regard, alternative feed ingredients have been used in animal feeding, including cassava root meal [6, 7] and sweet potato meal [8–12]. The use of cassava as an alternative to conventional energy feed stuffs like maize could help to reduce feed costs [13].

2. Cassava (*Manihot esculenta*) in chicken diet

Crops with starchy roots, tubers, rhizomes, corms, or stems are known as root and tuber crops. They are mostly utilized to make animal feed, human food (either raw or processed), starch, alcohol, and fermented drinks like beer. A root is an organ that grows from the root tissue and is a small, frequently enlarged storage organ with hairy stalks. A root is also a tuber. It grows from a rhizome, or a prolonged stem tissue; therefore, it is probably an enlarged storage organ. Consequently, a plant may also be a root, whereas a tuber is a root crop. For instance, potatoes, sweet potatoes, and yams are edible tubers, whereas carrots and cassava are root vegetables. There are differences between the growth patterns of edible tubers and edible root crops or plants. Since the plant's edible tubers and root vegetables are what fuel its above-ground growth, they are rich in starchy nutrients. While most vegetables grow above ground, root and tuber vegetables are the components of the plant that grow below the soil or on the soil surface. Rhizomes that grow underground can run parallel to or just below the soil's surface and can also run horizontally. Simply, a swelling portion of one of these rhizomes makes up the tuber. It may be possible to extract nutrients from these bloated chunks. In order for the plants to create healthy new growth in the spring, they want to store nutrients for them [14].

Cassava (*Manihot esculenta*) is grown for its underground starchy tuberous roots in tropical and subtropical regions. Over 800 million people worldwide eat mostly cassava roots, commonly known as cassava tubers [15]. Cassava roots are low in protein but abundant in calories because they are mostly constituted of starch and soluble carbohydrates. It is anticipated that more than 60% of the cassava grown in Africa will be consumed, with the other third being used to make secondary products [16]. The rising and expensive cost of feed ingredients has been a hindrance to global chicken production for many years. The majority of the time, different cereal crops are used as a conventional feed source for chickens. The competition between human and animal food and feed, as well as the use of these components in other industries, may be to blame for the ongoing increase in these feed ingredients. As a result, a substitute for the conventional energy and protein elements in chicken feed must be affordable and easily accessible and have an adequate nutritious composition and should have no discernible impact on chickens. Accordingly, Cassava (*Manihot esculenta*) is one such alternative. A common root tuber known as cassava is high in calcium, vitamins B and C, vital minerals, and carbohydrates, making it a viable substitute for maize in the diets of chickens. Cassava's low protein content, imbalanced amino acid profile, dustiness, and presence of anti-nutritional elements, however, limit its utilization in chicken diets. However, these flaws can be solved through proper processing and the inclusion of feed additives in the diet [17].

Different parts of cassava can be supplemented into diets of chickens. As a result, cassava tubers can be consumed boiled, mashed, deep-fried, and so on, and there are many food products based on cassava such as tapioca (cassava starch), a worldwide food ingredient; fufu (cassava flour boiled in water); and garri (fermented cassava mash), the two last popular foods in West and Central Africa [18]. The basic cassava products used in animal feeding are chips and pellets, which can partially or completely substitute the cereal grain in poultry feed [19]. The finger-like leaves, which are consumed as vegetables or used as animal feed, as well as different byproducts from the cassava processing industries, such as pomace and peels from starch, ethanol, and cassava food production, which can be used as livestock feed, are examples of other cassava products. Cassava flour, which should not be consumed by humans, can be used to make livestock feed [20]. Livestock feeding accounts for more than a third of the cassava crop's production [21]. Before being powdered or pelletized for use in commercial livestock feed, cassava tubers are first cut into slices and dried. Cassava chips can be made using basic household or village procedures as well as on a big, mechanized scale, and the processes utilized at various sizes of chip and pellet manufacturing are connected. The amount of cassava that needs to be processed, the cost of labor and capital, as well as the accessibility of relatively cheap energy, all influence the technology that is selected [22]. The availability of a better source of protein and the inclusion of enough methionine in the diet to meet both body requirements and cyanide detoxification are two factors that will determine whether utilizing cassava powder as a maize alternative in poultry feed is feasible [23]. While attempts have been made to reduce the dirtiness with the addition of oil and supplementation with suitable amounts of methionine and lysine amino acids, it has been beneficially observed that cassava root meal can replace up to 30% of maize in a broiler ration [24]. In a layer's diet, cassava root meal can completely replace maize [25]. Compared to chicken-fed rations with cassava chips or maize, broiler chicken-fed rations containing cassava pellets exhibited improved feed consumption [26]. For growing Japanese quail, about 35% cassava meal-based ration is suggested [27]. Cockerel starter birds can tolerate only about 28% level of cassava sievate in their ration [28]. Cassava root sievate and wet maize milling waste [29] can successfully replace maize by up to 35% without impacting the growth and feed utilization of finisher broilers.

The majority of research findings demonstrates that adding cassava to chickens' diets results in attractive responses in their behavior. For instance, substituting a 4:1 mixture of cassava root meal and leaves for maize in the diet of chickens can lower feed costs without affecting the layers' ability to gain weight or produce eggs [30]. Feeding broilers cassava chips supplemented with Moringa oleifera leaf meal at levels of 5 and 10% showed that cassava chips replacing maize at levels of 55.56% and 83.33%, respectively, in the diets had no negative effects on production and blood parameters [31]. Depending on dry matter consumption and growth performance of broilers, cassava root chips can completely substitute maize grain in broiler rations as an energy feed source [32]. On the other hand, based on the results of yields of major edible meat parameters, cassava root chips could replace maize grain by less than 50% in broilers diet, and 50% cassava root chips or 5% Moringa olifera meal, or a mixture of both can successfully be used in the ration of layers, substituting maize grain and soybean meal. The final body weight gain, total body weight gain, and daily body weight gain of broiler chickens were enhanced by substitution of noug seed cake with cassava leaf meal at a 4% substitution level. Thus, cassava leaves can be a good protein source to substitute the expensive Noug seed cake in the broiler ration [33].

3. Limitations on using cassava meal

The physical properties of cassava root meal, such as dustiness, poor pelleting quality, and poor pigmentation, tend to also limit the use of cassava as a feed ingredient in animal diets, in addition to the antinutritional factors and nutrient deficiencies inherent in raw and unprocessed cassava root. It has been noted that these physical restrictions, particularly in poultry, can lower feed intake and have an impact on body weight gain and feed conversion ratio. Animals fed a diet based on cassava that does not contain oil or that is fed as mash have also been shown to exhibit crop impaction and respiratory system irritation [34].

A cassava-based diet's dustiness is typically correlated with the form in which the feed is given to the animal. High levels of cassava meal in mash feed often produce dusty feed. Through proper pelleting, dustiness problems in cassava-based mash diets might be resolved, which would increase feed consumption and poultry performance. Pelletizing a cassava-based diet produces a diet that is denser, more homogeneous, and less dusty [35]. Diets based on cassava are about one-third less bulky after pelleting, which solves the dustiness problem. However, methods like the addition of oil or molasses can be used to address difficulties with dustiness in unpelleted chicken feed on farms without pelleting machinery.

To reduce dustiness, wet mashed feed can also be given to the birds; however, moist mashed feed should not be kept for an extended period of time to prevent contamination and deterioration. Lack of pigmentation is a further physical characteristic that restricts the use of cassava as a feed element for animals, particularly poultry. The color of cassava root meal is white (it does not have any pigmentation). It has been noted that feeding layers and broilers large amounts of cassava root meal causes pale meat and egg yolks, respectively. Due of minimal consumer appeal, it has been stated that these eggs and chicken flesh are sold for low prices.

When using a lot of cassava meal, the product quality should be improved by adding leaf meal or other pigmenting agents to the diet. Cassava root meal's lack of pigmentation can be avoided by adding at least 30–50 g of leaf meal per kg of poultry food. Leaf meals including those from young grass, ipil-ipil (leucaena) leaves, sweet potato leaves, and cassava leaves have been shown to be successful [35].

4. Methods to raise the nutritional value of cassava

Various processing techniques have been employed for many years to improve the nutritional content of cassava for use by humans and fowl. All of these techniques aim to remove different ANFs that are present in raw cassava, including hydrogen cyanide, phytate, saponin, and alkaloids [36]. These processing techniques have also been utilized to address physical issues like dustiness and a lack of pigmentation that tend to decrease performance and product quality and raise mortality when unprocessed cassava is used as a food or feed ingredient, as well as nutrient shortages. There are two types of cassava processing techniques: traditional [37] and modern [38]. Traditional cassava processing techniques include drying; boiling; parboiling/cooking; steaming; frying; roasting; addition of oil, molasses, and leaf meal; and utilization of natural fermentation processes in order to improve the nutritional composition and decrease the anti-nutrient content. HCN losses from these procedures range from 25 to 98% [39]. The addition of feed additives, such as nutrient supplements with amino acids, vitamins, and minerals; the addition of pigmentation agents; pelleting; the addition of synthetic enzymes; the microbial fermentation of cassava roots; and the genetic modification of the cassava plant are examples of contemporary methods of cassava processing [40].

5. Conclusion

Feed resources for chickens mainly come from activities directed toward human food production. The potential sources of feed for chicken production are mainly cereal crops and their byproducts. Shortage of conventional feed resources and continuous increment in its cost have been challenging chicken production in the tropics. Utilization of alternative feed resources is mandatory to overcome this challenge in the poultry sector. Root and tuber crops can be added in chicken diets. Cassava is among the root crops and can be included in chickens' ration. Thus, the tuber of cassava can be included in broilers diet and substitute up to 50% maize in the ration without harming the performances of both broiler and layer chickens. The use of cassava in chicken diet is limited by the presence of some anti nutritional factors. These factors can be reduced by different methods. In conclusion, cassava can be included in chickens' diet without causing negative impact on the performances of chickens.

Author details

Tagesse Tadesse
Department of Animal Science, College of Agricultural Sciences, Wachemo University, Hossana, Ethiopia

*Address all correspondence to: tagessetadesse76@gmail.com

References

[1] Dessie T, Ogle B. Village poultry production systems in the central highlands of Ethiopia. Tropical Animal Health and Production. 2001;**33**(6):521-537

[2] Solomon D. Growth performance of local and white Leghorn chicken under intensive management system. Journal Science. 2004;**27**:161-164

[3] Fafiolu AO, Oduguwa OO, CON I, Onwuka CFI, Adebule MA. Performance and egg quality assessment of laying hens fed malted sorghum sprouts base diets. In: Proceedings of the 9th Annual Conference of Animal Science Association of Nigeria. Abakaliki, Nigeria: Ebonyi State University; 2004. pp. 33-35

[4] Boufennara, Lopez S, Bousseboua H, Bodas R, Bouazza L. Chemical composition and digestibility of some browse plant species collected from Algerian arid rangelands. Spanish Journal of Agricultural Research. 2012;**10**(1):88-98

[5] Aregheore EM. Chemical composition and nutritive value of some tropical by-product feedstuffs for small ruminants—In vivo and in vitro digestibility. Animal Feed Science and Technology. 2000;**85**(1-2):99-109

[6] Sultana F, Khatun MF, Ali MA. Effect of dietary cassava tuber meal on egg production and egg quality of laying hen. Int J BioRes. 2011;**2**:1-4

[7] Sultan F, Ali MA, Jahan I. Growth performance meat yield and profitability of broiler chickens fed diets incorporating cassava tuber meal. Journal of Environmental Science & Natural Resources. 2012;**5**(1):47-53

[8] Agwunobi LN. Performance of broiler chickens fed sweet potato meal (*Ipomea batatas* L.) diets. Tropical Animal Health and Production. 1999;**31**:383-389

[9] Ayuk EA, Essien A. Growth and haematological response of broiler chicks fed graded levels of sweet potato (Ipomoea batata) meal as replacement for maize. International Journal of Poultry Science. 2009;**8**:485-488

[10] Beckford RC, Bartlett JR. Inclusion levels of sweet potato root meal in the diet of broilers I. Effect on performance, organ weights, and carcass quality. Poultry Science. 2015;**94**:1316-1322

[11] Maphosa T, Gunduza KT, Kusina J, Mutungamiri A. Evaluation of sweet potato tuber (*Ipomea batatas* l.) as a feed ingredient in broiler chicken diets. Livestock Research for Rural Development. 2003;**15**:Article #3

[12] Mozafari O, Ghazi S, Moeini MM. The effects of different levels of edible potat (Solanumtubresum) replacing maize on performance, serum metabolite and immune system of broiler chicks. Iranian Journal of animal Science. 2013;**3**(3):583-588

[13] Ukachukwu SN. Studies on the nutritive value of composite cassava pellets for poultry: Chemical composition and metabolizable energy. Livestock Research for Rural Development. 2005;**17**(11)

[14] Gardner K. The Difference between Tubers & Root Crops. 2019. Available from: https://www.gardenguides.com/13407021-thedifference-between-tubers-root-crops.html

[15] Ecocrop. Ecocrop Database. FAO;2011

[16] Scott G, Best R, Rosegrant MW, Bokanga M. Roots and tubers in the global food system—A vision statement to the year 2020. CIAT-CIP-IFPRI-IITA - IPGRICIP. 2000;**45**

[17] Ravindran V. Poulty Feed Availability and Nutrition in Developing Countries. Alternative Feedstuffs for Use in Poultry Feed Formulations. Poultry Development Review. Rome: FAO; 2013. pp. 60-63

[18] Kuiper L, Ekmecki B, Hamelink C, Hettinga W, Meyer S, Koop K. Bio-ethanol from cassava. In: Project Number: PBIONL062937. Ecofys Netherlands BV: Utrecht; 2007

[19] Iji PA, Bhuiyan MM, Chauynarong N, Barekatain MR, Widodo AP. Improving the nutritive value of alternative feed ingredients for poultry. In: Proceedings of the Recent Advanced in Animal Nutrition; Australia. 2011. pp. 115-122

[20] Boscolo WR, Hayashi C, Meurer F. Apparent digestibility of the energy and nutrients of conventional and alternative foods for Nile tilapia (Oreochromis niloticus). Revista Brasileira de Zootecnia. 2002;**31**(2):539-545

[21] FAO. FAOSTAT. Food and Agriculture Organization of the United Nations; 2011

[22] Hahn SK, Reynolds L, Egbunike GN. Cassava as livestock feed in Africa. In: Proc. IITA/ILCA/Univ. of Ibadan Workshop on the Potential Utilization of Cassava as Livestock Feed in Africa; 14-18 November 1988; Ibadan, Nigeria. 1992

[23] Okereke CO, Ukachukwu SN. Effect of dietary inclusion of composite cassava meal on egg production charateristics of laying hens [student thesis]. Micheal Okpara University. Department of Non Ruminant Animal Nutrution and Biochemistry; 2005. pp. 1-38

[24] Tekalegn Y, Etalem T, Getnet A. Poultry feed resources and coping mechanisms of challenges in Sidama zone, southern Ethiopia. 2017

[25] Akinola LAF, Oruwari BM. Response of laying hens total dietary replacement of maize with cassava. Nigerian Journal of Animal Production. 2007;**34**(2):196-202

[26] Bhuiyan MM, Iji PA. Energy value of cassava products in broiler chicken diets with or without enzyme supplementation. Asian-Australasian Journal of Animal Sciences. 2015;**28**(9):1317-1326

[27] Edache JA, Musa U, Karsin PD, Esilonu DO, Yisa A, Okpala EJ, et al. The feeding value of cassava meal diets for growing Japanese quail (Coturnix coturnix japonica). Nigerian Journal of Animal Production. 2007;**34**(1):77-82

[28] Nwokoro SO, Orheruata AM, Paul IO. Replacement of maize with cassava sievates in cockerel starter diets; some blood metabolic. In: Proceeding of 25th Annual Conference with NSAP. 2000. pp. 234-236

[29] Onwujiariri EB, Onyekwere MU, Okoronkwu MO, Okechukwu S. Evaluation of wet maize milling waste and cassava root sievaite as a replacement for maize in broiler finisher diets. In: Proc 42nd Annual Conference, Agricultural Society of Nigeria (ASA). October 19th-23rd 2008 Ebonyi State University Abakilike Nigeria. 2000. pp. 669-671

[30] Tewe OO, Bokanga M. Research Highlights Cassava Utilization, 2001. Ibadan, Nigeria: International Institute of Tropical Agriculture; 2001

[31] Olugbemi TS, Mutayoba SK, Lekule FP. Effect of Moringa (Moringa oleifera) inclusion in cassava based diets fed to broiler chickens. International Journal of Poultry Science. 2010;**9**:363-367

[32] Etalem T, Getachew A, Mengistu U, Tadelle D. Nutritional value of cassava root chips and Moringa Oleifera leaf meal in broiler and layer rations [PhD dissertation]. 2013

[33] Melesse A, Masebo M, Abebe A. The substitution effect of Noug seed (*Guizotia abyssinica*) cake with cassava leaf (*Manihot escutulata* C.) meal on feed intake, growth performance, and carcass traits in broiler chickens. Journal of Animal Husbandry and Dairy Science. 2018, 2018;**2**(2):1-9

[34] Okechuku SN. Studies on the nutritive value of composite cassava pellets for poultry: Chemical composition and metabolizable energy. Livestock Research for Rural Development. 2005;**17**:125. Available from: http://www.lrrd.org/lrrd17/11/ukac17125.htm

[35] Food and Agricultural Organisation (FAO). Better Farming Series 46: Use of Cassava and Sweet Potatoes in Animal Feeding. 1995. p. 47

[36] Kemdirim OC, Chukwu OA, Achinewhu SC. Effect of traditional processing of cassava on the cyanide content of gari and cassava flour. Plant Foods and Human Nutrition. 1995;**48**(4):335-339

[37] Padmaja G. Cyanide detoxification in cassava for food and feed uses. Critical Reviews in Food Science and Nutrition. 1995;**35**(4):299-339

[38] Siritunga D, Sayre R. Transgenic approaches for cyanogen reduction in cassava. Journal of Association of Official Analytical Chemists International. 2007;**90**(5):1450-1455

[39] Zvauya R, Muzondo MI. Reduction of cyanide levels in cassava during sequential sun drying and solid state fermentation. International Journal of Food Science and Nutrition. 1995;**46**(1):13-16

[40] Liu J, Zheng Q, Ma Q, Gadidasu KK, Zhang PJ. Cassava genetic transformation and its application in breeding. Integrated Plant Biology. 2011;**53**(7):552-569

Chapter 10

Advances in Cassava Trait Improvement and Processing Technologies for Food and Feed

Kariuki Samwel Muiruri and Anwar Aliya Fathima

Abstract

Cassava is an important staple crop globally; its roots and leaves are directly consumed as food or undergo secondary processing in food industries or as animal feed. Inherent biological challenges in cassava affect the quality and quantity of food and feed. Although trait consolidation has been explored, the diversity in cassava food products has led to stratification of target crop characteristics. Among the traits targeted, crop improvement for food includes: yield and starch quality for different applications such as pounding, resistant starch, waxy starch, and even post-harvest deterioration. The presence of the antinutritional compound cyanide reduces the quality of food and feed, and efforts to reduce cyanide levels are continuously explored. In this Chapter, we review biological and technological research efforts in cassava geared toward improving the quality/quantity of cassava for food and feed. These efforts cut across target trait improvement efforts to new bioprocessing technologies.

Keywords: food, feed, traits, cassava, agro-processing, improvement, starch, quality, quantity

1. Introduction

Cassava is a major staple food crop cultivated for its starchy roots and sometimes its leaves. It ranks among the top five crops as a food source and is consumed by over 700 million people globally [1]. Cassava not only plays a critical role in food security but also serves as a trading commodity in most rural areas where it is grown [2]. In most cassava growing regions, for example, cassava trade entails the sale of fresh or recently harvested roots. In the recent past, cassava has gained additional importance owing to its increased use in animal feeds, bio-energy, and processed food industry [3–5]. Cassava is increasingly being used to partially or fully replace grains in livestock [5] and chicken [6] feed without negative effects [7]. Cassava is better adapted to drought and reduced soil fertility compared to cereal crops (mostly maize) used in animal feeds [8]. Cassava, therefore, serves to a greater extent as an appropriate animal feed alternative during the unpredictable conditions being experienced with changes in climate. Cassava biomass is already being used for nonfood and nonfeed applications [9, 10] and the uses are still rising. The utility of cassava as food, feed, and agro-processing is challenged by intrinsic and extrinsic threats. Among many challenges, low productivity impacts cassava use, resulting from a wide range of factors including cultivar type [11], agricultural management practices [12], pests and diseases [13], and environmental conditions [14]. Post-harvest physiological deterioration (PPD), a loss in quality of harvested cassava roots due to physiological and biochemical processes, is another major challenge in cassava use as food and feed [15]. The PPD results in the darkening of harvested roots rendering them unusable as food and feed. Another major drawback in cassava utility is the presence of cyanide in the roots, which have antinutritional properties. Consumption of cyanide-laden cassava or constant exposure in a cassava processing industry can cause neurological disorders and even immediate death [16, 17]. Poor nutritive value of cassava roots is yet another challenge, particularly in areas where cassava is consumed regularly as a staple. Cassava roots are generally low in proteins [18] and minerals like iron and zinc [19]. These among other challenges compromise the quantity and quality of cassava produce and approach to overcome them have been extensively explored.

Cassava roots contain up to 80% starch on a dry-weight basis [20]. Starch polymers contain amylose and amylopectin, which are interconnected α-1,4 linear chains and β-1,6 branched chains of glucose monomers. Amylose is insoluble in water at low temperatures whereas amylopectin dissolves easily in water. The heating and cooling of native starch cause retrogradation, a condition in which the amylose and amylopectin chains realign to form a more crystalline structure. High amylose content in native starch can promote retrogradation [21] and subsequently, the starch polymer expels water, a process called syneresis, and becomes unfavorable for food applications [22]. Other constraints include water absorption and swelling behavior that influences the consistency of the product in certain baking applications and this is dependent on the amylose content, size, and structural integrity of starch granules [23].

Approaches aimed at overcoming challenges affecting cassava use as food and feed can be broadly grouped into two: 1) those that modify the biology of the plant with the ultimate goal of improving its performance in quality and quantity [24, 25]. These methods include conventional breeding and biotechnologies such as genetic engineering and genome editing. 2) Processing and bioprocessing methods that modify cassava product profiles [26, 27]. These include addition of missing but critical compounds that enhance cassava product quality or removal of unwanted and harmful metabolites. In addition, methods that change cassava products' profiles to fit intended use are grouped into this second category. This chapter reviews these different approaches used in addressing challenges affecting use of cassava and its products in food and feed industries.

2. Improvement of cassava crop for food and feed

2.1 Targeting cassava for yield improvement

The quantity of cassava harvested is important in both food and feed industries. Yield in cassava is influenced by, among other factors, genetics, environment, and the corresponding interactions [28]. The selection of cassava cultivars for increased yield breeding using crop physiology as the basis entails simple yield parameters that include harvest index (HI), biomass, and root dry matter content (DMC) [29]. In cassava, HI is the ratio of root wet weight to the total plant biomass whereas dry matter content is the percentage of fresh root weight that is dry matter. The higher variability in the fresh weight of the harvested cassava necessitated the use of HI as a cassava yield indicator. High levels of HI heritability have been observed even early on in cassava growth making it a crucial yield prediction trait in cassava [30]. A positive correlation between fresh root eating qualities and DMC exists [31]. Majority of processing in the cassava food industry involves drying of the roots. The DMC and HI equally determine a key target trait in improving cassava yield and have consistently been used in breeding programs in Brazil [11, 32], Uganda [33], Nigeria [34], and Thailand [35] among others.

2.2 Early bulking for faster harvest

The increase in cases of drought occasioned by climate change makes predictability of harvest time a herculean task. This in turn makes predictable food and feed production from cassava difficult, especially for varieties that take long time to mature [36]. Planting of early maturing cassava varieties is a clear strategy for adapting to climate change and ensuring a quick return on investment in both cassava food and feed industries. Early bulking (thickening of storage roots due to starch accumulation) is a genotype-dependent trait that can be used in breeding for quicker harvesting and, consequently, faster food provision [37]. Bulking in cassava is thought to occur when vegetative requirements of photoassimilates are less than the ones being generated [38]. Early bulking varieties are thought to be ready for harvest 6—8 months post-planting [39]. Early bulking varieties among pro-vitamin A cassava genotypes have been selected with the goal of ensuring early harvesting [40]. Early cassava bulking has also been achieved in efforts to breed drought-tolerant varieties [37, 41].

2.3 Crop improvement for increased levels of pro-vitamin A carotenoids

Vitamin A deficiency is common, particularly in areas where one type of food naturally low in pro-vitamin A is consumed as a staple. Most of the cassava varieties cultivated in regions where it is a staple food tend to be low in vitamin A [42]. Fortification of cassava is used as a strategy for enriching cassava produce, particularly flour with β-carotene [43, 44]. Fortification can be a costly option compared to biofortification, where cassava varieties naturally produce pro-vitamin A [45]. Cassava varieties able to accumulate pro-vitamin A carotenoids in the roots have been developed using biotechnology [45] and conventional approaches [46]. Overexpression of the deoxy-d-xylulose-5-phosphate synthase (DXS) and bacterial phytoene synthase (crtB) genes resulted in enhanced levels of β-carotene in the roots through biotechnology approaches. Breeding for improved β-carotene requires first screening for varieties with this trait and working upwards to enhance it [47]. Varieties with higher levels of β-carotene have been identified and bred into the major cassava lines [48–50]. In addition to β-carotene, varieties with increased pro-vitamin A have longer shelf life, albeit at the expense of low dry matter content [45].

2.4 Improvement of cassava against post-harvest physiological deterioration (PPD)

Post-harvest physiological deterioration results in cassava roots turning blue, black, or brown in color one to five days post-harvest. The discoloration is accompanied by a bitter taste rendering the roots unusable [51]. The PPD has been associated with both plant genetics as well as environment, particularly storage conditions. In Indonesia, different cassava genotypes were observed to have a range of 10 to more than 20% deterioration and were classified as having low, medium, and high susceptibility to PPD [51]. Similar variations have also been observed in cassava from Africa and South America [52–54]. The genes responsible for PPD and the associated quantitative trait loci (QTLs) have already been mapped in the cassava genome [55, 56]. In Nigeria, where extensive breeding for tolerance to PPD has been done, four major tolerance sources have been identified and include interspecific hybrids, gamma irradiated mutants where a gene involved in PPD is silenced, clones with high β-carotene, and an amylose-free waxy starch mutant [29, 57]. In addition to conventional breeding, genetic engineering strategies have also been applied in addressing the PPD challenge. Overexpression of reactive oxygen species (ROS) scavenging genes has shown promising results in reducing PPD [58]. One major challenge observed in silencing some of the genes associated with reduced PPD is loss in dry matter content and reduced starch storage [59]. Therefore, appropriate genes that minimize PPD while still retaining desirable traits like starch deposition and dry matter accumulation should be explored.

2.5 Improvement of cassava for specific starch profile

Cassava is prepared for food in different ways including boiling and pounding to make foods like "fufu" common in west Africa, and even in drinks, the most famous being bubble tea. The physicochemical qualities of cassava starch determine the culinary uses of the roots and associated products [60]. Cassava starch contains amylopectin and amylose, the former being longer and more branched and the latter linear. The levels of complexity in branching differentiate amylopectin's physical and chemical characteristics [61].

The variations in physical and chemical properties of starch in cassava are dependent on the genotype. There are four cassava genotype classes categorized on the basis of amylose levels: The waxy type contains a maximum of 2% amylose, semi-waxy with a maximum of 15%, normal-regular with a maximum of 35%, and high category with more than 35% amylose cumulative root starch yield [61]. Most varieties are grown for food fall within the normal-regular category. However, waxy varieties that completely lack amylose have been observed to exist as natural variants [62–64]. Naturally, waxy scratch variants have been observed to carry a mutation in the *granule-bound starch synthase (GBSS)* gene that renders it non-functionally [65]. Silencing of the GBSS gene through transgenic approaches has also resulted in waxy starch genotypes [66]. Cassava waxy starch is extensively used as a stabilizer in food storage because it does not have syneresis making it first among other starches in food stabilization.

Other than waxy starch, resistant starch, which is starch less amenable to enzymatic breakdown in the human stomach and reduced absorption in the small intestine, has been considered for a product profile [67]. The difficulty in breakdown of the RS bears resemblance to dietary fiber in being indigestible. The indigestibility of RS in the small intestine lowers the pH in the large intestine consequently increasing the fecal bulk, and consequently reducing the risks associated with cancer of the colon [68].

2.6 Developing cassava with reduced cyanide levels

The presence of cyanogenic glycosides in cassava roots risks its use for food and feed in both humans and animals. Development of cassava varieties with reduced cyanide levels is considered a viable approach to safe use of cassava root [69, 70]. To reduce the levels of cyanide in cassava a combination of both biotechnological and conventional approaches is being implemented. Through biotechnology, targeting cassava *CYP79D1* and *CYP79D2* genes resulted in a cyanogenic cassava, especially in the leaves [69, 70]. Low cyanogenic glucoside cassava varieties have been identified and selected for use in food production [71]. Combined, biotechnological, and conventional approaches to cyanide reduction can be used to enhance usability of both the raw and processed products from cassava roots for both food and feed (**Table 1**).

Target trait/ starch property	Biotechnologies/bioprocessing methods	Outcome	References
Starch content	ADP-glucose pyrophosphorylase overexpression	Increase in root tuber mass and enhanced starch content	[72]
Waxy starch	Conventional breeding	Low amylose, improved physico-chemical characteristics of starch	[64]
	CRISPR-Cas9 targeted mutagenesis of GBSSI	Low amylose	[66]
	Enzymatic modification of cassava starch using alpha-amylase/dry heat	Reduced amylose content	[73, 74]
Resistant starch	RNAi silencing of branching enzymes	High amylose	[75]
	Cross-linking with citric acid	Increased amylose content, altered crystallinity	[76]
Protein content	Expression of storage protein ASP1	Root tuber enhanced with amino acids	[77]
	Solid state fermentation (SSF) using Rhizopus sp., and soy protein fortification	Replacement for wheat flour	[78]
Carotenoids	Expression of deoxy-d-xylulose-5-phosphate synthase (DXS) and bacterial phytoene synthase (crtB)	Provitamin A production in roots	[45]
	Fermentation of cassava flour using Lactobacillus plantarum	Improved protein and Pro-vitamin A in cassava flour	[79]
Cyanide content	CRISPR-Cas9 mutagenesis of CYP79D1 and CYP79D2 genes	Reduction in cyanogenic glycosides	[70]
	Fermentation and wetting	Reduction in cyanide content in fermented leaves and roots	[80, 81]
Physicochemical properties of starch amylose, granular integrity	Expression of starch-phosphorylating enzyme glucan water dikinase (GWD) and silencing endogenous phosphatase and phosphogucan genes	Altered swelling and paste clarity, starch granule size	[82]
	Acid hydrolysis/enzymatic/high voltage electric treatment and phosphorylation	Decrease in amylose and starch content, low enthalpy of gelatinisation and damaged starch granule	[83, 84]
Retrogradation/ syneresis	Addition of hydrocolloids	Low syneresis, improved pasting viscosity	[85]
Post-harvest physiological deterioration (PPD)	RNAi mediated silencing of Feruloyl CoA 6′-hydroxylase gene	Reduction in PPD symptoms during storage	[86]

Table 1.
Summary of research on trait and bioprocessing approaches aimed at enhancing cassava quality and quantity for food and feed.

3. Cassava bioprocessing for improved food and feed quality

Cassava is a source of commonly used ingredients in the food industry, which, when processed, result in functionally unique value-added products [87]. Owing to the high carbohydrate content and gluten-free status of cassava products, their demand has increased. The increase in demand has necessitated the development of economically viable processing methods [19, 88]. The quality of processed cassava products is largely determined by the quality of starch. The starch quality is dependent on factors such as (i) method of extraction (ii) physicochemical characteristics and (iii) nutritive quality. Bioprocessing methods have been widely used to alter the properties of native cassava starch to be made suitable for the food and feed industry.

3.1 Improving cassava starch content

Conventionally, cassava starch is extracted from root tubers using maceration at optimum temperatures to produce pulp. To recover starch, the pulp containing 50–60% w/w of starch [89] is resuspended in 10 fold w/v of water followed by filtration, settling, decanting, and drying [90]. This method results in starch recovery t ranging from 18–25% per fresh cassava tuber weight. This makes the method laborious and time-consuming with minimal returns. Several modifications in the extraction methods have shown enhanced starch yield. The maceration of cassava tissues to produce pulp, for example, was improved by addition of microbial enzymes such as pectinases and polygalacturonase [91]. Recently, ultrasound-assisted extraction (UAE) was used to extract starch trapped inside cellulose fibers in cassava. The method produced the highest yield of 56.57% with a starch purity of 88.36% [92]. The limitations with starch losses in commercial-scale production units have been notably challenging and can be minimized by optimizing operation, design, and feed variables [93]. These processes have been further improved by addition of sulfur for improved starch granular stability and separation of the native starch from bound proteins and other impurities during the extraction process [94, 95].

3.2 Improving functional properties of cassava starch

The native starch can be processed to improve the physicochemical properties that maximize the use of cassava in the food and feed industry. The cassava processing methods involve either physical, chemical, enzymatic, or biological pre-treatments that modify starch content, nutrient content, texture, swelling power, solubility, pasting property, viscosity, gelatinization, crystallinity, and anti-nutrient content of the modified starch [96].

Physical modification of starch includes thermal or nonthermal methods. Thermal modifications such as pre-gelatinization, heat and moist treatment, dry heating, annealing, and microwaving involve the treatment of starch at various temperatures and pressures [97]. These methods have been very effective in deformation of granular structure of starch and in reducing crystallinity. Nonthermal modifications such as ball milling, cold plasma technology, hydrostatic pressure, microfluidization, pulse electric field, and ultrasound require no heat treatment, and, therefore, reduce energy consumption [98]. These modifications have shown the ability to maintain the granular integrity and increase surface activity of starch consequently improving pasting and swelling properties of cassava starch [98]. Recently, a combination of methods utilizing dry heat and ozone treatment has been shown to affect starch molecule size, structural properties, and gelatinization [99].

The commonly used chemical treatments include acid hydrolysis, cross-linking, and oxidation [100]. The functional groups in chemicals react with hydroxyl groups in native starch to produce modified starch. The modification of cassava starch at various concentrations of hydrochloric acid and temperatures has been shown to alter the fraction of amylose and amylopectin and crystallinity of native starch [101]. Acid hydrolysis improves water holding capacity and water absorption, which can alter pasting properties of starch [102]. The treatment of native starch with weak organic acids such as citric acid is advantageous in culinary applications because it reportedly improves granular starch yield and avoids depolymerization. Citric acid has also been used for cross-linking of native starch [103]. Cross-linking introduces covalent interactions and reinforces the hydrogen bonds between starch molecules and prevents movement of polymer chains thereby reducing retrogradation and providing resistance to shearing and thermal decomposition during storage [104]. Moreover, natural organic acids such as lactic acid, malic acid, and citric acid are safe for consumption and generally have applications in the food industry as acidity regulators, flavoring agents, etc. Some of the studies highlighted here have highlighted a combination of modifications that improve properties of native starch [105]. Post-acid treatment of lactic acid hydrolyzed native starch by drying under UV irradiation resulted in the depolymerization of native starch and improved its baking expansion [96]. Lactic acid hydrolysis combined with microwave heating or fermentation and esterification in the presence of ethanol have shown better properties of modified starch for use in food coatings [106]. Other cross-linking agents that are commonly used are adipic and acetic acid mixed anhydrides, phosphorus oxychloride, sodium trimetaphosphate, sodium hydroxide, and ethylene glycol diacrylate epichlorohydrin (EPI) [107–109]. Acetylation and esterification improved water retention and reduced retrogradation tendency in cassava starch compared to sorghum and potato starch [110, 111].

The starch processing approaches discussed have their pros and cons. Chemical processes, for example, can be high starch-yielding but are prone to residual by-products of the chemical reactions. Physical processes are considered clean and sustainable in comparison with chemical starch processing/modifying approaches. However, physical modification techniques involving prolonged heating may reduce starch viscosity and stability and are moderately effective in reducing the amylose content making the starch less suitable for baking [112]. Alternative enzymatic treatments and fermentation methods of starch modification alter amylose content by addition of hydrolyzing enzymes or treatment with microorganisms that can break down amylose [113]. Retting is a traditional process that allows microbial activity on plant materials to dissolve cell wall polysaccharides such as cellulose and pectin by immersion in water [114]. The pasting properties can be influenced by the combination of microbes used for retting [115]. Starch modifying and starch converting enzymes mainly perform hydrolysis, transglycosylation, or cleavage of α-1, 4 linked glucan and α-1, 6 linked branches, extensively reviewed for their use in baking applications by Mikolo et al. [116].

3.3 Bioprocessing to improve nutritive quality of cassava starch

The consumption of starch containing high amounts of amylose (resistant starch) has several health benefits [117]. Like previously noted, resistant starch (RS) escapes digestion in the small intestine and is fermented by the gut microbiota in the large intestine [118]. The fermentation end-products such as short-chain fatty acids, acetic acid, and butyric acid have been found to reduce the effects of chronic inflammatory health conditions and also improve growth of beneficiary gut microbes [119]. Chemical modification of cassava starch using octenyl succinic anhydride has been shown to improve resistant starch content in cassava [120]. The most common and highly resistant starch-yielding methods use enzymatic debranching of native starch using isoamylase or pullulanase followed by autoclaving and/or high-pressure annealing [121, 122].

Cassava has a high energy content and is widely regarded as a good source of dietary fiber, minerals, and vitamins. However, the protein content of cassava roots is significantly low compared to cassava leaves [123]. Pre-processing of cassava roots through drying and fermentation improved protein, fiber, and shelf-life of cassava products [124]. Co-fermentation with legumes has been shown to enrich protein content in cassava [125]. Alternatively, cassava-based industries have developed high-quality cassava flour (HQCF) from processed cassava roots without fermentation, no-off odor, and appealing taste [126]. HCQF can replace wheat flour by addition of protein-rich mushroom flour to HQCF, which makes it a suitable food commodity [127, 128].

Research has been focused on developing different processing methods to reduce cyanide content [129]. The use of cassava as food and feed is primarily limited by high cyanide content and antinutrients that reduce nutrient availability as discussed in the previous sections [130, 131]. The edible parts of cassava including root tubers and leaves have been found to contain toxic cyanogenic glycosides (CNglcs) namely linamarin and lotaustralin [132]. Prolonged consumption of cassava can increase risk of cyanide poisoning in humans and animals [133]. During tissue damage, the released glycosides come in contact with enzymes (linamarase) and form acetone cyanohydrin, a less stable intermediate that is either spontaneously or enzymatically (hydroxynitrile lyase) converted to volatile and toxic hydrogen cyanides (HCN) [134]. Previously, processing techniques such as pounding cassava tubers and soaking the paste in cold water, wilting, drying, boiling, ensiling, and fermentation removed most of the cyanides [135, 136]. During cassava root retting, microbial strains have been found to naturally evolve with ability to produce enzymes that degrade CNglcs and have been proven to reduce cyanides efficiently during the process of fermentation [137–139]. Alternatively, the application of enzymes that degrade plant cell wall polysaccharides such as cellulases and hemicellulases may indirectly trigger release of linamarin due to tissue damage caused by these enzymes [140]. In another study, cassava leaves were washed, dried, and treated with bicarbonate to efficiently reduce the cyanide contents [141]. Bicarbonate treatment was most efficient in reducing cyanide levels in cassava leaves in comparison to thermal, enzymatic, and ultrasonic methods but significantly reduced nutritive components such as ascorbic acid [142]. The other antinutrients present in cassava leaves including polyphenols, nitrates, and phytates significantly reduce absorption of proteins and essential minerals [143, 144]. Fermentation of cassava leaves has been the most promising method to reduce antinutrients and concomitantly retain the nutritive value of cassava and flavors [145, 146]. Other potential methods include boiling, steaming, dry-roasting, and microwaving [147].

4. Conclusion

Cassava has been identified as an exemplary crop for developing a sustainable food system due to its hardiness, relatively better adaptation to abiotic stress, particularly drought and higher productivity. Challenges inherent in the crop have necessitated crop improvement and bioprocessing efforts to improve its use for food, feed, and bioenergy sectors. Some of these methods as highlighted in this chapter have progressively been implemented to enhance the quality of cassava food products in terms of palatability and dietary benefits and ensure transformation of cassava as a potential raw material to produce economic and high-value livestock and poultry feed. Crop improvement approaches as summarized in this chapter have been employed to produce varieties with increased essential micro-nutrients and pro-vitamin A to address malnutrition and adopt cassava as a biofortified crop. Protein supplementation of cassava flour has promoted cassava as an ideal choice for high-quality diet that can replace conventional food crops. The utility of cassava has been undoubtedly increased through modern technologies making cassava versatile to serve nontraditional subsistence roles with enhanced market value.

Author details

Kariuki Samwel Muiruri[1*] and Anwar Aliya Fathima[2]

1 Kenyatta University, School of Plant Sciences, Nairobi, Kenya

2 Department of Bioinformatics, Saveetha School of Engineering, Saveetha Institute of Medical and Technical Sciences, Chennai, India

*Address all correspondence to: muiruri.samwel@ku.ac.ke

References

[1] FAOSTAT [Internet]. [cited 2022 Nov 10]. Available from: https://www.fao.org/faostat/en/

[2] Rozi F, Krisdiana R, Sutrisno I. Pattern of cassava demand as the promising commodity in the future. In: 1st International Conference on Sustainable Agricultural Socio-economics, Agribusiness, and Rural Development (ICSASARD 2021). France: Atlantis Press; 2021. pp. 31-36

[3] Li S, Cui Y, Zhou Y, Luo Z, Liu J, Zhao M. The industrial applications of cassava: Current status, opportunities and prospects. Journal of the Science of Food and Agriculture. 2017 Jun;**97**(8):2282-2290

[4] Yulianto A, National Laboratory for Starch Technology, Agency for the Assesment and Application of Technology, Lampung, Indonesia, Kayati FN, et al. The effect of extrusion temperature on changes of characteristics of partially pregelatinized Cassava starch. International Journal of Chemical Engineering and Applications. 2020;**11**:67-70. DOI: 10.18178/ijcea.2020.11.2.782

[5] Aso SN, Teixeira AA, Achinewhu SC. Cassava residues could provide sustainable bioenergy for cassava producing nations. In: Cassava. London: IntechOpen. 2018;**2018**:219. DOI: 10.5772/intechopen.72166

[6] Zheng Y, Zhao Y, Xue S, Wang W, Wang Y, Cao Z, et al. Feeding value assessment of substituting Cassava (*Manihot esculenta*) residue for concentrate of dairy cows using an in vitro gas test. Animals. 2021;**11**:307. DOI: 10.3390/ani11020307

[7] Chang EP, Abdallh ME, Ahiwe EU, Mbaga S, Zhu ZY, Fru-Nji F, et al. Replacement value of cassava for maize in broiler chicken diets supplemented with enzymes. Journal of Animal Science. 2020;**33**(7):1126-1137

[8] Muiruri SK, Ntui VO, Tripathi L, Tripathi JN. Mechanisms and approaches towards enhanced drought tolerance in cassava (*Manihot esculenta*). Current Plant Biology. 2021;**28**:100227. DOI: 10.1016/j.cpb.2021.100227

[9] Okudoh V, Schmidt S, Trois C. Biogas Production in Africa: Benefit Potentials of Cassava Biomass. London: LAP Lambert Academic Publishing; 2015. p. 84

[10] Nuwamanya E, Chiwona-Karltun L, Kawuki RS, Baguma Y. Bio-ethanol production from non-food parts of cassava (Manihot esculenta Crantz). Ambio. 2012;**41**(3):262-270

[11] de Andrade LR, Sousa MBE, Oliveira EJ, de Resende MD, Azevedo CF. Cassava yield traits predicted by genomic selection methods. PLoS One 2019;**14**(11):e0224920

[12] Kintché K, Hauser S, Mahungu NM, Ndonda A, Lukombo S, Nhamo N, et al. Cassava yield loss in farmer fields was mainly caused by low soil fertility and suboptimal management practices in two provinces of the Democratic Republic of Congo. European Journal of Agronomy. 2017;**89**:107-123. DOI: 10.1016/j.eja.2017.06.011

[13] Uke A, Tokunaga H, Utsumi Y, Vu NA, Nhan PT, Srean P, et al. Cassava mosaic disease and its management in Southeast Asia. Plant Molecular Biology. 2022;**109**(3):301-311

[14] Yabuta S, Fukuta T, Tamaru S, Goto K, Nakao Y, Khanthavong P, et al. The Productivity of Cassava (*Manihot esculenta* Crantz) in Kagoshima, Japan,

Which Belongs to the Temperate Zone. Agronomy. 2021;**11**:2021. DOI: 10.3390/agronomy11102021

[15] Djabou ASM, Carvalho LJCB, Li QX, Niemenak N, Chen S. Cassava postharvest physiological deterioration: A complex phenomenon involving calcium signaling, reactive oxygen species and programmed cell death. Acta Physiologiae Plantarum. 2017;**39**(4):91

[16] Dhas PK, Chitra P, Jayakumar S, Mary AR. Study of the effects of hydrogen cyanide exposure in cassava workers. Indian Journal of Occupational Environmental Medicine. 2011;**15**(3):133-136

[17] Akintonwa A, Tunwashe OL. Fatal cyanide poisoning from cassava-based meal. Human & Experimental Toxicology. 1992;**11**(1):47-49

[18] Stephenson K, Amthor R, Mallowa S, Nungo R, Maziya-Dixon B, Gichuki S, et al. Consuming cassava as a staple food places children 2-5 years old at risk for inadequate protein intake, an observational study in Kenya and Nigeria. Nutrition Journal. 2010;**9**:9

[19] Morgan NK, Choct M. Cassava: Nutrient composition and nutritive value in poultry diets. Animal Nutrition. 2016;**2**(4):253-261

[20] Olomo V, Ajibola O. Processing factors affecting the yield and physicochemical properties of starch from cassava chips and flour. Starch/Stärke. 2003;**55**(10):476-481. DOI: 10.1002/star.200300201

[21] Biduski, Bárbara, Wyller Max Ferreria da Silva, Rosana de Colussi, Shanise Lisie, Loong-Tak Lim, Álvaro Renato Guerra Dias, and Elessandra Rosa Zavareze. "Starch hydrogels: The influence of the amylose content and gelatinization method." International Journal of Biological Macromolecules. 2018;**113**:443-449

[22] Mizrahi S. Syneresis in food gels and its implications for food quality. In: Chemical Deterioration and Physical Instability of Food and Beverages. Sawston, CA: Woodhead Publishing; 2010. pp. 324-348

[23] Arya S, Sadawarte P, Ashish W. Importance of damaged starch in bakery products-a review. 2019. Retrieved April 1 2015

[24] Mbanjo EGN, Rabbi IY, Ferguson ME, Kayondo SI, Eng NH, Tripathi L, et al. Technological innovations for improving cassava production in sub-Saharan Africa. Frontiers in Genetics. 2020;**11**:623736

[25] Fathima AA, Sanitha M, Tripathi L, Muiruri S. Cassava (*Manihot esculenta*) dual use for food and bioenergy: A review. Food and Energy Security. 2022;**12**(1):e380. DOI: 10.1002/fes3.380

[26] Omede AA, Ahiwe EU, Zhu ZY, Fru-Nji F, Iji PA. Improving cassava quality for poultry feeding through application of biotechnology. Cassava. 2018:214-264. DOI: 10.5772/intechopen.72236

[27] Padmaja G. Cyanide detoxification in cassava for food and feed uses. Critical Reviews in Food Science and Nutrition. 1995;**35**(4):299-339

[28] Phoncharoen P, Banterng P, Vorasoot N, Jogloy S, Theerakulpisut P, Hoogenboom G. Growth rates and yields of cassava at different planting dates in a tropical Savanna climate. Scientia Agricola. 2019;**76**:376-388. DOI: 10.1590/1678-992x-2017-0413

[29] Ceballos H, Hershey C, Iglesias C, Zhang X. Fifty years of a public cassava

breeding program: Evolution of breeding objectives, methods, and decision-making processes. Theoretical and Applied Genetics. 2021;**134**(8):2335-2353

[30] Kawano K, Narintaraporn K, Narintaraporn P, Sarakarn S, Limsila A, Limsila J, et al. Yield improvement in a multistage breeding program for Cassava. Crop Science. 1998;**38**:325-332. DOI: 10.2135/cropsci1998.0011183x003800020007x

[31] Kawano K, Fukuda WMG, Cenpukdee U. Genetic and environmental effects on dry matter content of cassava root. Crop Science. 1987;**27**:69-74. DOI: 10.2135/cropsci1987.0011183x002700010018x

[32] Sagrilo E, Filho PSV, Pequeno MG, Gonçalves-Vidigal MC, Kvitschal MV. Dry matter production and distribution in three cassava (Manihot esculenta Crantz) cultivars during the second vegetative plant cycle. Brazilian Archives of Biology and Technology. 2008;**51**:1079-1087. DOI: 10.1590/s1516-89132008000600001

[33] Manze F, Rubaihayo P, Ozimati A, Gibson P, Esuma W, Bua A, et al. Genetic gains for yield and virus disease resistance of cassava varieties developed over the last eight decades in Uganda. Frontiers in Plant Science. 2021;**21**(12):651992

[34] Jiwuba L, Danquah A, Asante I, Blay E, Onyeka J, Danquah E, et al. Genotype by environment interaction on resistance to cassava green mite associated traits and effects on yield performance of cassava genotypes in Nigeria. Frontiers in Plant Science. 2020;**11**:572200

[35] Maraphum K, Saengprachatanarug K, Wongpichet S, Phuphuphud A, Sirisomboon P, Posom J. Modified specific gravity method for estimation of starch content and dry matter in cassava. Heliyon. 2021;**7**(7):e07450

[36] Okogbenin E, Fregene M. Genetic analysis and QTL mapping of early root bulking in an F1 population of non-inbred parents in cassava (Manihot esculenta Crantz). Theoretical and Applied Genetics. 2002;**106**(1):58-66

[37] Chipeta MM, Shanahan P, Melis R, Sibiya J, Benesi IRM. Early storage root bulking index and agronomic traits associated with early bulking in cassava. Field Crops Research. 2016;**198**:171-178. DOI: 10.1016/j.fcr.2016.09.004

[38] el Sharkawy MA. Cassava biology and physiology. Plant Molecular Biology. 2003;**53**:621-641. DOI: 10.1023/b:plan.0000019109.01740.c6

[39] Amelework AB, Bairu MW. Advances in genetic analysis and breeding of cassava (Crantz): A review. Plants. 2022;**20**:12. DOI: 10.3390/plants11121617

[40] David Badewa O, Gana Saba A, Kolo Tsado E, Dele TK. Selection of early bulking performance among pro vitamin A cassava genotypes based on selective indices of fresh storage root yield and harvest index. International Journal of Genetic Genome. 2020;**8**(1):11

[41] Campos H, Caligari PDS. Genetic Improvement of Tropical Crops. Newyork: Springer; 2017. p. 320

[42] Belalcazar J, Dufour D, Andersson MS, Pizarro M, Luna J, Londoño L, et al. High-throughput phenotyping and improvements in breeding cassava for increased carotenoids in the roots. Crop Science. 2016;**56**:2916-2925. DOI: 10.2135/cropsci2015.11.0701

[43] Mosha TC, Laswai HS, Mtebe K, Paulo AB. Control of vitamin A

deficiency disorders through fortification of cassava flour with red palm oil: A case study of Kigoma district, Tanzania. Ecology of Food and Nutrition. 1998;**37**:569-593. DOI: 10.1080/03670244.1998.9991566

[44] Wang S. The alleviation of vitamin a deficiency through staple food fortification in ghana. Review of Agricultural and Applied. 2018;**21**:94-102. DOI: 10.15414/raae.2018.21.02.94-102

[45] Beyene G, Solomon FR, Chauhan RD, Gaitán-Solis E, Narayanan N, Gehan J, et al. Provitamin A biofortification of cassava enhances shelf life but reduces dry matter content of storage roots due to altered carbon partitioning into starch. Plant Biotechnology Journal. 2018;**16**(6):1186-1200

[46] Oluba OM, Oredokun-Lache AB, Odutuga AA. Effect of vitamin A biofortification on the nutritional composition of cassava flour (gari) and evaluation of its glycemic index in healthy adults. Journal of Food Biochemistry. 2018;**42**:e12450. DOI: 10.1111/jfbc.12450

[47] Njoku DN, Vernon G, Egesi CN, Asante I, Offei SK, Okogbenin E, et al. Breeding for enhanced β-carotene content in cassava: Constraints and accomplishments. Journal of Crop Improvement. 2011;**25**:560-571. DOI: 10.1080/15427528.2011.594978

[48] Nduwumuremyi A, Melis R, Shanahan P, Theodore A. Genetic inheritance of pulp colour and selected traits of cassava (Manihot esculenta Crantz) at early generation selection. Journal of the Science of Food and Agriculture. 2018;**98**(8):3190-3197. DOI: 10.1002/jsfa.8825

[49] Moorthy SN, Jos JS, Nair RB, Sreekumari MT. Variability of β-carotene content in cassava germplasm. Food Chemistry. 1990;**36**:233-236. DOI: 10.1016/0308-8146(90)90058-c

[50] Athanase N, Rob M. Gene action and heterosis in F1 clonal progenies of cassava for β-Carotene and farmers' preferred traits. Heliyon. 2019;**5**:e01807. DOI: 10.1016/j.heliyon.2019.e01807

[51] Rachmawati RS, Khumaida N, Ardie SW, Sukma D, Sudarsono S. Effects of harvest period, storage, and genotype on postharvest physiological deterioration responses in cassava. Biodiversitas Journal of Biological Diversity. 2021;**23**(1):100-109. DOI: 10.13057/biodiv/d230113

[52] Venturini MT, Santos LR, Vildoso CIA, Santos VS, Oliveira EJ. Variation in cassava germplasm for tolerance to post-harvest physiological deterioration. Genetics and Molecular Research. 2016;**15**(2). DOI: 10.4238/gmr.15027818

[53] Tumuhimbise R, Melis R, Shanahan P. Genetic variation in cassava for postharvest physiological deterioration. Archives of Agronomy and Soil Science. 2015;**61**:1333-1342. DOI: 10.1080/03650340.2014.995641

[54] Moyib KO, Mkumbira J, Odunola OA, Dixon AG, Akoroda MO, Kulakow P. Genetic variation of postharvest physiological deterioration susceptibility in a cassava germplasm. Crop Science. 2015;**55**:2701-2711. DOI: 10.2135/cropsci2014.11.0749

[55] Reilly K, Bernal D, Cortés DF, Gómez-Vásquez R, Tohme J, Beeching JR. Towards identifying the full set of genes expressed during cassava post-harvest physiological deterioration. Plant Molecular Biology. 2007;**64**:187-203. DOI: 10.1007/s11103-007-9144-0

[56] Beeching JR, Han Y, Gómez-Vásquez R, Day RC, Cooper RM. Wound

and Defense responses in cassava as related to post-harvest physiological deterioration. Phytochemical Signals and Plant—Microbe Interactions. 1998;**1998**:231-248. DOI: 10.1007/978-1-4615-5329-8_12

[57] Chávez AL, Sánchez T, Ceballos H, Rodriguez-Amaya DB, Nestel P, Tohme J, et al. Retention of carotenoids in cassava roots submitted to different processing methods. Journal of the Science of Food and Agriculture. 2007;**87**:388-393. DOI: 10.1002/jsfa.2704

[58] Xu J, Duan X, Yang J, Beeching JR, Zhang P. Enhanced reactive oxygen species scavenging by overproduction of superoxide dismutase and catalase delays postharvest physiological deterioration of cassava storage roots. Plant Physiology. 2013;**161**(3):1517-1528

[59] Beyene G, Chauhan RD, Gehan J, Siritunga D, Taylor N. Cassava shrunken-2 homolog MeAPL3 determines storage root starch and dry matter content and modulates storage root postharvest physiological deterioration. Plant Molecular Biology. 2022;**109**(3):283-299

[60] Nuwamanya E, Baguma Y, Kawuki RS, Rubaihayo PR. Quantification of starch physicochemical characteristics in a cassava segregating population. African Crop Science Journal. 2010;**16**(3):191-202. DOI: 10.4314/acsj.v16i3.54380

[61] Chisenga SM. Primary quality control parameters for cassava raw materials. Cassava. 2021:151-167. DOI: 10.5772/intechopen.97879

[62] do Carmo CD, MBE S, Dos Santos Silva PP, Oliveira GAF, Ceballos H, de Oliveira EJ. Identification and validation of mutation points associated with waxy phenotype in cassava. BMC Plant Biology. 2020;**20**(1):164

[63] do Carmo CD, Sousa MB, Brito AC, de Oliveira EJ. Genome-wide association studies for waxy starch in cassava. Euphytica. 2020;**216**(5). DOI: 10.1007/s10681-020-02615-9

[64] Toae R, Sriroth K, Rojanaridpiched C, Vichukit V, Chotineeranat S, Wansuksri R, et al. Outstanding characteristics of Thai Non-GM bred waxy cassava starches compared with normal cassava starch, waxy cereal starches and stabilized cassava starches. Plants. 2019;**24**(8):11. DOI: 10.3390/plants8110447

[65] Ceballos H, Sánchez T, Morante N, Fregene M, Dufour D, Smith AM, et al. Discovery of an amylose-free starch mutant in cassava (Manihot esculenta Crantz). Journal of Agricultural and Food Chemistry. 2007;**55**(18):7469-7476

[66] Bull SE, Seung D, Chanez C, Mehta D, Kuon JE, Truernit E, et al. Accelerated ex situ breeding of - and -edited cassava for modified starch. Science Advances. 2018;**4**(9):eaat6086

[67] Sánchez T, Dufour D, Moreno IX, Ceballos H. Comparison of pasting and gel stabilities of waxy and normal starches from potato, maize, and rice with those of a novel waxy cassava starch under thermal, chemical, and mechanical stress. Journal of Agricultural and Food Chemistry. 2010;**58**(8):5093-5099

[68] Charles AL, Chang YH, Ko WC, Sriroth K, Huang TC. Influence of amylopectin structure and amylose content on the gelling properties of five cultivars of cassava starches. Journal of Agricultural and Food Chemistry. 2005;**53**(7):2717-2725

[69] Jørgensen K, Bak S, Busk PK, Sørensen C, Olsen CE, Puonti-Kaerlas J, et al. Cassava plants with a depleted

cyanogenic glucoside content in leaves and tubers. Distribution of cyanogenic glucosides, their site of synthesis and transport, and blockage of the biosynthesis by RNA interference technology. Plant Physiology. 2005;**139**(1):363-374

[70] Juma BS, Mukami A, Mweu C, Ngugi MP, Mbinda W. Targeted mutagenesis of the CYP79D1 gene via CRISPR/Cas9-mediated genome editing results in lower levels of cyanide in cassava. Frontiers in Plant Science. 2022;**13**:1009860. DOI: 10.3389/fpls.2022.1009860

[71] Mbah EU, Nwankwo BC, Njoku DN, Gore MA. Genotypic evaluation of twenty-eight high- and low-cyanide cassava in low-land tropics, Southeast Nigeria. Heliyon. 2019;**5**(6):e01855

[72] Ihemere U, Arias-Garzon D, Lawrence S, Sayre R. Genetic modification of cassava for enhanced starch production. Plant Biotechnology Journal. 2006;**4**(4):453-465

[73] Boonna S, Rolland-Sabaté A, Lourdin D, Tongta S. Macromolecular characteristics and fine structure of amylomaltase-treated cassava starch. Carbohydrate Polymers. 2019;**205**:143-150

[74] Chandanasree D, Gul K, Riar CS. Effect of hydrocolloids and dry heat modification on physicochemical, thermal, pasting and morphological characteristics of cassava (*Manihot esculenta*) starch. Food Hydrocolloids. 2016;**52**:175-182. DOI: 10.1016/j.foodhyd.2015.06.024

[75] Zhou W, Zhao S, He S, Ma Q, Lu X, Hao X, et al. Production of very-high-amylose cassava by post-transcriptional silencing of branching enzyme genes. Journal of Integrative Plant Biology. 2020;**62**(6):832-846

[76] Remya R, Jyothi AN, Sreekumar J. Effect of chemical modification with citric acid on the physicochemical properties and resistant starch formation in different starches. Carbohydrate Polymers. 2018;**202**:29-38

[77] Zhang P, Jaynes JM, Potrykus I, Gruissem W, Puonti-Kaerlas J. Transfer and expression of an artificial storage protein (ASP1) gene in cassava (Manihot esculenta Crantz). Transgenic Research. 2003;**12**(2):243-250

[78] Begum R, Rakshit SK, Mahfuzur Rahman SM. Protein fortification and use of cassava flour for bread formulation [internet]. International Journal of Food Properties. 2011;**14**:185-198. DOI: 10.1080/10942910903160406

[79] Rosales-Soto MU, Gray PM, Fellman JK, Scott Mattinson D, Ünlü G, Huber K, et al. Microbiological and physico-chemical analysis of fermented protein-fortified cassava (*Manihot esculenta* Crantz) flour. LWT - Food Science and Technology. 2016;**66**:355-360. DOI: 10.1016/j.lwt.2015.10.053

[80] Kobawila SC, Keleke S. Reduction of the cyanide content during fermentation of cassava roots and leaves to produce bikedi and ntoba mbodi, two food products from Congo. African Journal of Biotechnology. 2005;**4**:689-696. DOI: 10.5897/ajb2005.000-3128

[81] Bradbury JH, Cliff J, Denton IC. Uptake of wetting method in Africa to reduce cyanide poisoning and konzo from cassava. Food and Chemical Toxicology. 2011;**49**(3):539-542

[82] Wang W, Hostettler CE, Damberger FF, Kossmann J, Lloyd JR, Zeeman SC. Modification of cassava root starch phosphorylation enhances starch functional Properties. Frontiers in Plant Science. 2018;**9**:1562

[83] Ačkar Đ, Grec M, Grgić I, Gryszkin A, Styczyńska M, Jozinović A, et al. Physical properties of starches modified by phosphorylation and high-voltage electrical discharge (HVED). Polymers. 2022;**14**(16):3359. DOI: 10.3390/polym14163359

[84] Ahmed AS, Igbo UE, Igwe CC. Evaluation of the physico-chemical properties of acid thinned cassava starch. Nigerian Food Journal. 2005;**23**(1):85-90. DOI: 10.4314/nifoj.v23i1.33603

[85] Leite TD, Nicoleti JF, Penna ALB, Franco CML. Effect of addition of different hydrocolloids on pasting, thermal, and rheological properties of cassava starch. Food Science and Technology. 2012;**32**:579-587. DOI: 10.1590/s0101-20612012005000074

[86] LiuS,ZainuddinIM,VanderschurenH, Doughty J, Beeching JR. RNAi inhibition of feruloyl CoA 6′-hydroxylase reduces scopoletin biosynthesis and post-harvest physiological deterioration in cassava (Manihot esculenta Crantz) storage roots. Plant Molecular Biology. 2017;**94**(1-2):185-195

[87] Udoro EO, Anyasi TA, Jideani AIO. Process-induced modifications on quality attributes of cassava (*Manihot esculenta* Crantz) flour. Processes. 2021;**9**:1891. DOI: 10.3390/pr9111891

[88] Garske RP, Mercali GD, Thys RCS, Cladera-Olivera F. Cassava starch and chickpea flour pre-treated by microwave as a substitute for gluten-free bread additives. Journal of Food Science and Technology. 18 Sep 2022;**60**(1):53-63. DOI: 10.1007/s13197-022-05586-y

[89] Bunterngsook B, Laothanachareon T, Natrchalayuth S, Lertphanich S, Fujii T, Inoue H, et al. Optimization of a minimal synergistic enzyme system for hydrolysis of raw cassava pulp. RSC Advances. 2017;7:48444-48453. DOI: 10.1039/c7ra08472b

[90] Kaur K, Ahluwalia P, Singh H. Cassava: Extraction of starch and utilization of flour in bakery products. International Journal of Food Fermentation Technology. 2016;**6**(2):351

[91] Martos MA, Zubreski ER, Combina M, Garro OA, Hours RA. Isolation of a yeast strain able to produce a polygalacturonase with maceration activity of cassava roots. Food Science and Technology. 2013;**33**:332-338

[92] Setyaningsih W, Karmila FRN, Cahyanto MN. Process optimization for ultrasound-assisted starch production from cassava (Manihot esculenta Crantz) using response surface methodology. Agronomy. 2021;**11**(1):117

[93] Saengchan K, Nopharatana M, Songkasiri W. Recovery of tapioca starch from pulp in a conical basket centrifuge–effects of rotational speed and liquid to solid (L/S) ratio on cake formation and starch–pulp separation efficiency. Separation and Purification Technology. 2014;**127**:192-201

[94] Lim ST, Lee JH, Shin DH, Lim HS. Comparison of protein extraction solutions for rice starch isolation and effects of residual protein content on starch pasting properties. Starch. 1999;**51**:120-125. DOI: 10.1002/(sici)1521-379x(199904)51:4<120::aid-star120>3.0.co;2-a

[95] Sriroth K, Wanlapatit S, Piyachomkwan K, Oates CG. Improved cassava starch granule stability in the presence of sulphur dioxide. Starch. 1998;**50**:466-479. DOI:10.1002/(sici)1521-379x(199812)50:11/12<466::aid-star466>3.0.co;2-0

[96] Sumardiono S, Jos B, Pudjihastuti I, Sari RJ, Kumala WDN, Cahyono H. Effect of chemical modification, drying

method, and drying temperature on baking expansion and the physicochemical properties of cassava starch. Journal of Food Processing and Preservation. Jan 2022;**46**(1). DOI: 10.1111/jfpp.16111

[97] Che LM, Li D, Wang LJ, Chen XD, Mao ZH. Micronization and hydrophobic modification of cassava starch. International Journal of Food Properties. 2007;**10**:527-536. DOI: 10.1080/10942910600932982

[98] Maniglia BC, Castanha N, Rojas ML, Augusto PED. Emerging technologies to enhance starch performance. Current Opinion in Food Science. 2021;**37**:26-36. DOI: 10.1016/j.cofs.2020.09.003

[99] Lima DC, Maniglia BC, Matta Junior MD, Le-Bail P, Le-Bail A, Augusto PED. Dual-process of starch modification: Combining ozone and dry heating treatments to modify cassava starch structure and functionality. International Journal of Biological Macromolecules. 2021;**167**:894-905

[100] Wang Z, Mhaske P, Farahnaky A, Kasapis S, Majzoobi M. Cassava starch: Chemical modification and its impact on functional properties and digestibility, a review. Food Hydrocolloids. 2022;**129**:107542. DOI: 10.1016/j.foodhyd.2022.107542

[101] Beninca C, Colman TAD, Lacerda LG, Filho MAS, Bannach G, Schnitzler E. The thermal, rheological and structural properties of cassava starch granules modified with hydrochloric acid at different temperatures. Thermochimica Acta. 2013;**552**:65-69. DOI: 10.1016/j.tca.2012.10.020

[102] Babu AS, Parimalavalli R, Jagannadham K, Rao JS. Chemical and structural properties of sweet potato starch treated with organic and inorganic acid. Journal of Food Science and Technology. 2015;**52**(9):5745-5753

[103] Utomo P, Nizardo NM, Saepudin E. Crosslink modification of tapioca starch with citric acid as a functional food. In: Proceedings of the 5th International Symposium on Current Progress in Mathematics and Sciences (ISCPMS2019). Newyork: AIP publishing; 2020. DOI: 10.1063/5.0010364

[104] Shah N, Mewada RK, Mehta T. Crosslinking of starch and its effect on viscosity behaviour. Reviews in Chemical Engineering. 2016;**32**(2):265-270. DOI: 10.1515/revce-2015-0047

[105] Khurshida S, Das MJ, Deka SC, Sit N. Effect of dual modification sequence on physicochemical, pasting, rheological and digestibility properties of cassava starch modified by acetic acid and ultrasound. International Journal of Biological Macromolecules. 2021;**188**:649-656

[106] Sumardiono S, Pudjihastuti I, Jos B, Taufani M, Yahya F. Modification of cassava starch using combination process lactic acid hydrolysis and micro wave heating to increase coated peanut expansion quality. In: AIP Conference Proceedings. AIP Publishing LLC; 24 May 2017;**1840**(1):060005. DOI: 10.1063/1.4982285

[107] Marques PT, Pérégo C, Le Meins JF, Borsali R, Soldi V. Study of gelatinization process and viscoelastic properties of cassava starch: Effect of sodium hydroxide and ethylene glycol diacrylate as cross-linking agent. Carbohydrate Polymers. 2006;**66**:396-407. DOI: 10.1016/j.carbpol.2006.03.028

[108] Méndez PA, Méndez ÁM, Martínez LN, Vargas B, López BL. Cassava and banana starch modified with maleic anhydride-poly (ethylene glycol) methyl

ether (Ma-mPEG): A comparative study of their physicochemical properties as coatings. International Journal of Biological Macromolecules. 2022;**205**:1-14

[109] Jyothi AN, Moorthy SN, Rajasekharan KN. Effect of cross-linking with epichlorohydrin on the properties of cassava (*Manihot esculenta* Crantz) Starch. Starch. 2006;**58**:292-299. DOI: 10.1002/star.200500468

[110] Palavecino PM, Penci MC, Ribotta PD. Effect of sustainable chemical modifications on pasting and gel properties of sorghum and cassava starch. Food and Bioprocess Technology. 2020;**13**:112-120. DOI: 10.1007/s11947-019-02381-0

[111] Mbougueng PD, Tenin D, Scher J, Tchiégang C. Influence of acetylation on physicochemical, functional and thermal properties of potato and cassava starches. Journal of Food Engineering. 2012;**108**:320-326. DOI: 10.1016/j.jfoodeng.2011.08.006

[112] Lin CL, Lin JH, Lin JJ, Chang YH. Properties of high-swelling native starch treated by heat–moisture treatment with different holding times and iterations. Molecules. 2020;**25**:5528. DOI: 10.3390/molecules25235528

[113] Dolas KA, Ranveer RC, Tapre AR, Nandane AS, Sahoo AK. Effect of starch modification on physico-chemical, functional and structural characterization of cassava starch (*Manihot esculenta* Crantz). Food Research. 2020;**4**:1265-1271. DOI: 10.26656/fr.2017.4(4).075

[114] Ngolong Ngea GL, Guillon F, Essia Ngang JJ, Bonnin E, Bouchet B, Saulnier L. Modification of cell wall polysaccharides during retting of cassava roots. Food Chemistry. 2016;**213**: 402-409

[115] Ze NN, Ndjouenkeu R, Ngang JJE. Effect of accelerated retting process on physiochemical and pasting Properties of cassava (Manihot esculenta Crantz) flours. Asian Food Science Journal. 2020;**2020**:45-52. DOI: 10.9734/afsj/2020/v15i430160

[116] Park SH, Na Y, Kim J, Kang SD, Park KH. Properties and applications of starch modifying enzymes for use in the baking industry. Food Science and Biotechnology. 2018;**27**(2):299-312

[117] Shi YC, Maningat CC. Resistant Starch: Sources, Applications and Health Benefits. New Jersey: John Wiley & Sons; 2013. p. 312

[118] Tan FPY, Beltranena E, Zijlstra RT. Resistant starch: Implications of dietary inclusion on gut health and growth in pigs: A review. Journal of Animal Science and Biotechnology. 2021;**12**(1):124

[119] Jia L, Dong X, Li X, Jia R, Zhang HL. Benefits of resistant starch type 2 for patients with end-stage renal disease under maintenance hemodialysis: A systematic review and meta-analysis. International Journal of Medical Sciences. 2021;**18**(3):811-820

[120] Zhang B, Mei JQ, Chen B, Chen HQ. Digestibility, physicochemical and structural properties of octenyl succinic anhydride-modified cassava starches with different degree of substitution. Food Chemistry. 2017;**229**:136-141

[121] Abioye VF, Adeyemi IA, Akinwande BA, Kulakow P, Maziya-Dixon B. Effect of autoclaving on the formation of resistant starch from two Nigeria Cassava (*Manihot esculenta*) varieties. Food Research. 2018;**2**:468-473. DOI: 10.26656/fr.2017.2(5).205

[122] Lertwanawatana P, Frazier RA, Niranjan K. High pressure intensification

of cassava resistant starch (RS3) yields. Food Chemistry. 2015;**181**:85-93

[123] Montagnac JA, Davis CR, Tanumihardjo SA. Nutritional value of cassava for use as a staple food and recent advances for improvement. Comprehensive Reviews in Food Science and Food Safety. 2009;**8**(3):181-194

[124] Onyango SO, Abong GO, Okoth MW, Kilalo DC, Mwang'ombe AW. Effect of pre-treatment and processing on nutritional composition of cassava roots, millet, and cowpea leaves flours. Frontiers in Sustainable Food Systems. 2 Jun 2021;**5**:625735. DOI: 10.3389/fsufs.2021.625735

[125] Agbon CA, Ngozi EO, Onabanjo OO. Production and nutrient composition of Fufu made from a mixture of cassava and cowpea flours. Journal of Culinary Science & Technology. 2010;**8**:147-157. DOI: 10.1080/15428052.2010.511096

[126] TECA, FAO. Production of high quality cassava flour. Accessed December 9, 2022. https://teca.apps.fao.org/teca/fr/technologies/4574

[127] Ekunseitan OF, Obadina AO, Sobukola OP, Omemu AM, Adegunwa MO, Kajihausa OE, et al. Nutritional composition, functional and pasting properties of wheat, mushroom, and high quality cassava composite flour. Journal of Food Processing and Preservation. 2017;**41**:e13150. DOI: 10.1111/jfpp.13150

[128] Ayandipe DO, Adebowale AA, Obadina O, Sanwo K, Kosoko SB, Omohimi CI. Optimization of high-quality cassava and coconut composite flour combination as filler in chicken sausages. Journal of Culinary Science & Technology. 2022;**20**:1-32. DOI: 10.1080/15428052.2020.1799280

[129] Montagnac JA, Davis CR, Tanumihardjo SA. Processing techniques to reduce toxicity and Antinutrients of cassava for use as a staple food. Comprehensive Reviews in Food Science and Food Safety. 2009;**8**:17-27. DOI: 10.1111/j.1541-4337.2008.00064.x

[130] Ravindran V. Cassava leaves as animal feed: Potential and limitations. Journal of the Science of Food and Agriculture. 1993;**61**:141-150. DOI: 10.1002/jsfa.2740610202

[131] Quinn AA, Myrans H, Gleadow RM. Cyanide content of cassava food products available in Australia. Foods. 11 May 2022;**11**(10):1384. DOI: 10.3390/foods11101384

[132] Jørgensen K, Morant AV, Morant M, Jensen NB, Olsen CE, Kannangara R, et al. Biosynthesis of the cyanogenic glucosides linamarin and lotaustralin in cassava: Isolation, biochemical characterization, and expression pattern of CYP71E7, the oxime-metabolizing cytochrome P450 enzyme. Plant Physiology. 2011;**155**(1):282-292

[133] Cressey P, Reeve J. Metabolism of cyanogenic glycosides: A review. Food and Chemical Toxicology. 2019;**125**:225-232

[134] White WLB, Arias-Garzon DI, McMahon JM, Sayre RT. Cyanogenesis in cassava. The role of hydroxynitrile lyase in root cyanide production. Plant Physiology. 1998;**116**(4):1219-1225

[135] Tweyongyere R, Katongole I. Cyanogenic potential of cassava peels and their detoxification for utilization as livestock feed. Veterinary and Human Toxicology. 2002;**44**(6):366-369

[136] Nambisan B. Strategies for elimination of cyanogens from cassava

for reducing toxicity and improving food safety. Food and Chemical Toxicology: An International Journal Published for the British Industrial Biological Research Association. 2011;**49**(3):690-693

[137] Obi CN, Okezie O, Ukaegbu T. Fermentation reduces cyanide content during the production of cassava flours from sweet and bitter cassava tuber varieties. Asian Food Science Journal. 2019;**2019**:1-10. DOI: 10.9734/afsj/2019/v11i130050

[138] Iwuoha CI, Banigo EOI, Okwelum FC. Cyanide content and sensory quality of cassava (*Manihot esculenta* Crantz) root tuber flour as affected by processing. Food Chemistry. 1997;**58**:285-288. DOI: 10.1016/0308-8146(95)00184-0

[139] Nkoudou NZ, Essia JJN. Cyanides reduction and pasting Properties of cassava (Manihot Esculenta Crantz) flour as affected by fermentation process [internet]. Food and Nutrition Sciences. 2017;**08**:326-333. DOI: 10.4236/fns.2017.83022

[140] Sornyotha S, Kyu KL, Ratanakhanokchai K. An efficient treatment for detoxification process of cassava starch by plant cell wall-degrading enzymes. Journal of Bioscience and Bioengineering. 2010;**109**(1):9-14

[141] Mikolo B, Nakavoua AHW, Nkounga CK. Effects of cassava leaves detoxification processes on the physicochemical and sensory qualities of *Saka Saka*. American Journal of Applied Chemistry. 2021;**9**:109. DOI: 10.11648/j.ajac.20210904.13

[142] Zhong Y, Xu T, Ji S, Wu X, Zhao T, Li S, et al. Effect of ultrasonic pretreatment on eliminating cyanogenic glycosides and hydrogen cyanide in cassava. Ultrasonics Sonochemistry. 2021;**78**:105742

[143] Wobeto C, Corrêa AD, de Abreu CMP, dos Santos CD, Pereira HV. Antinutrients in the cassava (Manihot esculenta Crantz) leaf powder at three ages of the plant. Ciência e Tecnologia de Alimentos. 2007;**27**:108-112. DOI: 10.1590/s0101-20612007000100019

[144] Delimont NM, Haub MD, Lindshield BL. The impact of tannin consumption on iron bioavailability and status: A narrative review. Current Developmental Nutrition. 2017;**1**(2):1-12

[145] Hawashi M, Altway A, Widjaja T, Gunawan S. Optimization of process conditions for tannin content reduction in cassava leaves during solid state fermentation using. Heliyon. 2019;**5**(8):e02298

[146] Terefe ZK, Omwamba M, Nduko JM. Effect of microbial fermentation on nutritional and antinutritional contents of cassava leaf. Journal of Food Safety. 2022;**42**(3). DOI: 10.1111/jfs.12969

[147] Ekpo UA, Baridia DF. Effect of processing on the chemical and antinutritional properties of cassava leaves (sweet and bitter varieties). ARC Journal of Nutrition and Growth. 2020;**6**(2):6-12

Chapter 11

The Cyanogenic Potential of Certain Cassava Varieties in Uganda and Their Fermentation-Based Detoxification

Benson Oloya, Christopher Adaku and Morgan Andama

Abstract

Cassava is the leading staple food in the developing world, providing an essential diet for about half a billion individuals. However, cassava contains significantly toxic compounds, the cyanogenic glycosides. Ingestion of such toxins in large quantities can lead to acute cyanide poisoning and may cause death in both humans and animals. Therefore, cassava may present a potential health risk to consumers. Information regarding the cyanogenic glycoside content is vital in averting health risks associated with cassava consumption. Accordingly, the seven most common local cultivars in Zombo district and six improved cultivars were grown and later characterized based on their cyanogenic potential. Additionally, the root tubers of *Nyar-udota* and *Nyar-papoga* were fermented to detoxify them from the cyanogens. The cyanogenic glycoside levels in the selected cultivars surpassed the critical value of 10 ppm established by the World Health Organization. The improved cassava had lower and moderately identical concentrations of HCN, unlike the local varieties. Cyanogenic contents were highest at 8-10 months. Fermentation led to substantial detoxification of the cyanogens, and the decrease varied with the fermentation period. In making choices for the cultivation and consumption of cassava, it is crucial to consider the cultivar, period of harvesting, and detoxification by fermentation.

Keywords: cassava, cyanogenic potential, cyanogenic glycosides, cyanide poisoning, detoxification, fermentation, food safety, food security

1. Introduction

Cassava produced by 105 countries is the basic food for more than 600 million people worldwide [1]. Cassava is a very important food source in the tropics, ranking third after rice and maize [2, 3]. It is presently one of Uganda's most vital food crops, ranking second to bananas in terms of the area it occupies, per capita consumption, and total production [1]. About 275 million tons of cassava were produced globally in 2018, with the largest producer being Africa (contributing 61.1% of the total), followed by Asia (29.0%), the Americas (9.8%), and Oceania (0.1%) [4]. In 2020, cassava production globally exceeded 302 million tons, with more than half of the production recorded in Africa [5]. Nigeria is the world's leading cassava producer, producing 35 million metric tons. In comparison, Uganda's cassava production is about 5 million metric tons each year, and the traditional cassava growing regions in Uganda include the North, West Nile, and Eastern parts of the country [1].

Cassava, a carbohydrate-rich crop, has many uses, including food for human consumption, animal feeds, fuel for producing biofuel & ethanol, and industrial raw material in making paper, citric acid, clothing, alcohol, medicine, and chemicals [4]. Cassava is easily grown and can produce better yields in good and even poor soils, subject to dry conditions. The roots are starchy and may be sweet or bitter, and the young leaves are a good source of protein [6]. Owing to the perceived agricultural advantages of cassava growing and the increasing demand for food as a result of population pressures, cassava usage has been extending to some parts of Africa and elsewhere where it was not formerly used [7]. Traditionally, cassava has been grown as a food security crop, a form of protection against drought and the failure of other staple crops. It is mainly planted in the first rainy season of the year rather than the second, and it is customarily intercropped with beans, maize, and sweet potatoes [8].

Africa's cassava production is mainly for domestic consumption. The cassava supports local food security as well as an economic activity for the farmers, with the main products being fresh cassava roots and processed cassava products [4]. In Uganda, cassava growing is mostly practiced by smallholder farmers covering 1–2 acres of land to ensure food security and generate income. However, there is an effort by the government of Uganda to encourage large-scale cassava production to cater for the ever-growing commercial uses of cassava in the baking industry, pharmaceutical industries, and the manufacture of paper board and starch, biofuel, and alcohol [1]. Most of the cassava is sold as dry cassava chips or cassava flour milled from the dried chips and as fresh cassava roots, especially in urban areas [1].

Unfortunately, all the cassava cultivars produce toxic compounds in the form of cyanogenic glycosides, such as linamarin and lotaustralin, in varying concentrations, ranging from around 10 mg/kg to over 500 mg/kg fresh weight basis [9]. The cyanogenic glycoside content in cassava roots is determined by the cultivar and the growth conditions [10]. These cyanogens are spread in all parts of the plant, with the highest amounts in the leaves and the root cortex (skin layer). The root parenchyma (interior) has comparatively smaller amounts of cyanogens. The so-called sweet cassava varieties have only a small amount of cyanogens in the parenchyma so that after peeling, these roots can be safely boiled and eaten [6]. Bitter cassava's bitter taste is primarily due to linamarin [11]. Cassava produces the two cyanogenic glycosides as a defence mechanism to prevent predator attacks.

The cyanogenic glycosides are nitrile-containing plant secondary compounds that produce cyanide (cyanogenesis) after their enzymatic breakdown. A cyanogenic glucoside is typically a D-glucose joined by a β-linkage to an acetone cyanohydrin derivative [12]. There are about 25 different types of cyanogenic glucosides; the only difference between them is the residual group attached to the end of the acetone cyanohydrin. Linamarin has a hydrogen atom, whilst lotaustralin has a methyl ($-CH_3$) group.

Cassava produces the cyanogenic glucosides in a particular way. The first step is the conversion of L-valine into (*Z*)-2-methylpropanal oxime, which is catalysed by two similar cytochromes (P450s) which are encoded by the genes CYP79D1/D2 [13]. There are two of these genes since *M. esculenta* is an allopolyploid. In the next step (*Z*)-2-methylpropanal oxime reacts to acetone cyanohydrin, and in the final step, a glucose molecule is bound to the acetone derivative, forming linamarin.

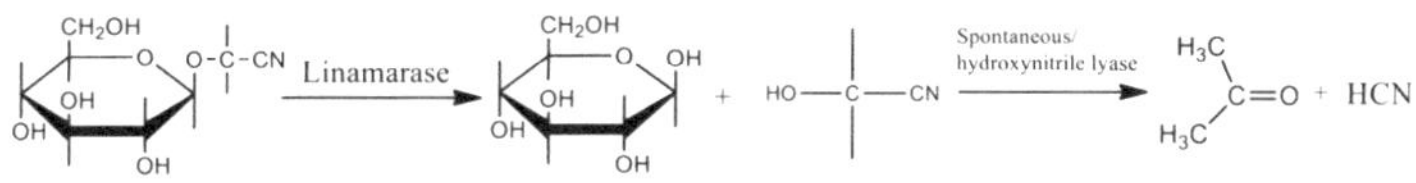

Figure 1.
Cyanogenesis in cassava.

The physiology and the biochemistry of cyanogenesis (**Figure 1**) in cassava have been well studied [14, 15]. Cyanogenesis in cassava starts when there is damage in the plant tissue. When the vacuole is raptured, linamarin is released, and it is hydrolyzed by a cell wall-associated β-glycosidase, linamarase [14]. Linamarin hydrolyzes producing an unstable hydroxynitrile intermediate, acetone cyanohydrin, and glucose. Acetone cyanohydrin spontaneously decomposes to form acetone and HCN at pH >5.0 or temperatures >35°C. Acetone cyanohydrin can also be broken down by the enzyme hydroxynitrile lyase (HNL) [16–20].

In Africa, consumption of poorly processed cassava, specifically by nutritionally compromised individuals, has led to several cyanide-associated health disorders [21]. The severity of these disorders is dependent on the quantity of the cyanogens consumed, the frequency of cyanogen exposure, and the consumer's health. The presence of toxins in cassava presents a health risk because inadequate preparation of cassava can leave sufficient quantities of residual cyanide in cassava products. The consumption of cassava and its products containing significant amounts of cyanogens causes cyanide poisoning with symptoms of dizziness, vomiting, headache, stomach pains, diarrhoea, weakness, nausea, and occasionally death [22, 23]. The HCN is very poisonous because it binds to the Fe^{2+} in haemoglobin, forming cyanohaemoglobin [24]. As a result, there is impediment of the respiratory cycle because the binding affinity of cyanide is much higher than the equivalent binding affinity of oxygen.

Ingestion of cyanide from cassava aggravates goitre and cretinism in areas deficient in iodine [25] and is almost undoubtedly the cause of konzo in central, eastern, and southern Africa. Konzo is an irreversible paralysis of the legs that starts suddenly, occurring mostly in children and women of childbearing age [26–28]. Tropical ataxic neuropathy (TAN) is a chronic condition of gradual onset and occurs in older people who consume a monotonous cassava diet. It causes loss of vision, deafness, weakness, and ataxia of gait [29–31].

The body's major defence in countering cyanide's toxic effects is converting it to thiocyanate mediated by the enzyme rhodanese [32]. Therefore, individuals with low protein and in particular low cysteine intake in their diets are more vulnerable to cyanide poisoning since detoxifying cyanide to thiocyanate by rhodanese requires cysteine as a substrate [32]. In addition, a number of minor reactions help in the detoxification of ingested cyanide. Firstly, cystine can react directly with the cyanide forming 2-imino-thiazolidine-4-carboxylic acid, which is excreted in the saliva and urine [33]. Secondly, a small amount of the cyanide may be converted into formic acid and then excreted in urine [33]. Thirdly, cyanide can react with hydroxycobalamine (vitamin B12) to form cyanocobalamine, which is excreted in the urine and bile. Reabsorption of cyanocobalamine may also occur by the intrinsic factor mechanism in the ileum, permitting effective recirculation of vitamin B12 [33]. Fourthly, methaemoglobin effectively competes with cytochrome oxidase for cyanide, and its formation from haemoglobin, effected by sodium nitrile or amylnitrite, is exploited in the treatment of cyanide intoxication [33].

In Uganda and the West Nile sub-region in particular, excessive consumption of bitter cassava is responsible for disability amongst children. The region depends on cassava as its primary food source. Dr. Tito Beyeza, an orthopaedic surgeon at Makerere University College of Health Sciences, reported that 10 out of 40 children who underwent surgery at Nebbi Hospital had cyanide, signifying a serious health hazard. He added that removing cyanide requires surgery, which is expensive. 'A surgery like this is ordinarily done in Mulago at Uganda Shillings 800,000', he said. The regional coordinator of the Uganda Society for Disabled Children, Mr. Stephen Eguma, also noted that several families are affected. 'It is an expensive disease to treat for our poor parents here. And they let the children just grow with the deformity and disability which affects the child's future' [34].

Nevertheless, in an attempt to deter thieves, animals, and pests, many farmers from cassava-growing countries oftentimes prefer the bitter varieties [35]. In some places, the more-toxic cassava varieties are a fallback resource (a 'food security crop') during famine [36]. Generally, higher cyanide content correlates with higher yields. During drought times, the cyanide content of both sweet and bitter cassava varieties rises [37]. Bitter cassava varieties are more readily available and cheaper during drought periods because they are more drought resistant. However, due to food shortages during drought, less time is sometimes available for the complete processing required, leaving sufficient quantities of the cyanogens in cassava [38].

Substantial reduction in the per capita cyanide intake could prevent the medical conditions caused by cyanide overload. It is, therefore, crucial to characterise cassava cultivars based on their cyanogenic potential so that cultivars with the lowest levels of toxins are recommended for household consumption. Also, to realise the full potential value of cassava, a lot has to be done at the processing level. Therefore, better and more effective processing methods, especially fermentation [39], have to be promoted and improved to reduce further the cyanide content in cassava flour to within acceptable limits (safe level) of 10 ppm, set by the World Health Organisation (WHO) [36].

2. Materials and methods

2.1 Materials

The materials used during this research included containers (basins), airtight polythene bags, a thermometer, a refrigerator, a distillation flask, a kitchen knife, a pH meter. Others included a reciprocating shaker, filter funnels, micro burette, 125 mL Erlenmeyer flasks, Filter paper (Whatman #42), disposable plastic vials, and distillation apparatus. The main reagents that were used during laboratory analysis were concentrated sulphuric acid, sodium hydroxide, potassium permanganate, 5% potassium iodide solution, 0.02 N silver nitrate, potassium dichromate, ferrous ammonium sulphate, and distilled water.

2.2 Sample acquisition

The cassava varieties used were obtained from the same garden in Agure village, Palei-west ward, Zombo Town council in Zombo district, Uganda.

2.2.1 Cultivation of cassava

A plot of land measuring 20 × 7 m was cleared manually, tilled, and 13 ridges measuring 18 × 0.6 m were made. The spacing between the ridges was 0.5 m. Stems of six improved cultivars of cassava (NASE 03, NASE 09, NASE 14, NASE

19, TME 14, and TME 204) were obtained from National Agricultural Research Organisation (NARO) at Abii Farm, Arua district, whilst the seven local cassava cultivars (‘*Bisimwenge*’, ‘*Nyar-anderian*’, ‘*Nyar-papoga*’, ‘*Nyar-pamitu*’, ‘*Nyar-matia*’, ‘*Nyar-udota*’, and ‘*Terengule*’) were collected from local peasant farmers in Zombo district, Uganda.

The cuttings from each cultivar, measuring about 27 cm in length, were planted at about 45° on the crest of the ridges [40]. Care was taken to ensure that the buds were not inverted during planting in order to prevent delayed sprouting [40]. The planting distance was about 0.5 × 0.5 m. Weeding was done at 4, 8, and 12 weeks, respectively, after planting since the crop was planted as a sole crop [41].

2.2.2 Collection and preparation of cassava samples

For the determination of cyanogenic glycosides content variation with cassava age, the samples were collected and prepared monthly (on the 15th of each month) for cassava aged 7–13 months. For the comparative analysis of the cyanogenic glycosides content in the various cassava cultivars, the samples were obtained only at the age of 13 months.

Fresh cassava root samples were obtained directly from the garden using a hand hoe. The soils were removed and then the samples were transported home in polythene bags for preparation and then to the Government Analytical Laboratory for analysis. 40 g of each peeled and washed sample was mashed using a wooden pestle and mortar and weighed. The samples were then kept in a deep freezer at a temperature of - 4°C awaiting analysis within 24 hours.

For heap fermentation, cassava roots grown for thirteen (13) months were got from the garden and the peels were removed using a kitchen knife. The peeled root tubers were subjected to partial sun drying at a temperature range of 28 to 40°C and at varied periods (0, 1, 2, 3, and 4 hours). The dried root tubers were heaped together on dry banana leaves with polythene sheets underneath and then covered with dry banana leaves, followed by black polythene sheets. The root tubers were heaped to enable terrestrial fermentation by the growth of moulds, and it was carried out in a grass-thatched hut having a clay floor to afford steady warmth. The period of fermentation was varied by withdrawing some of the heaped cassava after 2, 3, 4, 5, and 7 days for *Nyar-papoga*, and 0, 2, 3, 4, 6, 8, and 10 days for *Nyar-udota* variety.

The moulds from the fermented cassava roots were removed by scrapping them with a blunt kitchen knife. The cassava was pounded and then subjected to sun-drying for around 8 hours at a temperature ranging from 28 to 40°C. The dried cassava was then milled and analysed at the Government Analytical Laboratory. As a control, fresh tubers that were not fermented but sun-dried as well as a fresh tuber that was not dried, were milled and analysed.

2.3 Determination of level of cyanides in cassava

The standard method of FAO [42] was used to analyse the cassava samples at the Government Analytical Laboratory (GE058/07) in Kampala. Briefly, in order to set free all the bound hydrocyanic acid, the sample (10 to 20 g) was placed in a distillation flask, and distilled water (about 200 ml) was added and left to stand for two to 4 hours. The mixture was distilled with steam and 150–200 ml of distillate was collected in a solution of 0.5 g of sodium hydroxide in 20 ml of water. The distillate was then diluted to a volume of 250 ml.

To 100 ml of distillate was added 8 ml of 5% potassium iodide solution and titrated with 0.02 N silver nitrate (1 ml of 0.02 N silver nitrate corresponds to 1.08 mg of hydrocyanic acid) using a micro burette. The endpoint was shown by a faint but permanent turbidity, which was easily recognised, particularly against a black background. When all the cyanide ions have reacted with the silver ions, any excess silver ions react with the iodide ions giving a precipitate of silver iodide.

$$HCN_{(aq)} + AgNO_{3(aq)} \rightarrow HNO_{3(aq)} + AgCN_{(aq)} \quad (1)$$

$$Ag^{+}_{(aq)} + I^{-}_{(aq)} \rightarrow AgI_{(s)} \quad (2)$$

2.4 Data analysis

Graphs were generated using computer packages: SPSS 16 and Microsoft Excel from the results of laboratory analysis. Descriptive statistics for the overall HCN levels in each of the two cassava varieties (local and improved) were obtained. Student t test was used to compare the amount of HCN in the local and improved cassava varieties. Results were significant at 0.05 level. The experimental data were analysed using the two-way ANOVA for comparison of the effect of the duration of fermentation (days) and period of partial drying on the amount of hydrogen cyanide. The variation of HCN level with the duration of fermentation in *Nyar-udota* cassava cultivar was analysed using One-way ANOVA.

3. Results and discussions

3.1 Effect of age of cassava on the levels of hydrogen cyanide

The effect of the age of cassava on the levels of hydrogen cyanide was studied, and the result showing the trend is shown in **Figure 2**.

The level of hydrogen cyanide generally showed an increasing pattern from the 8th month up to the 10th month for varieties (NASE 9, TME 14, *Nyar-anderiano,* and *Bisimwenge*). Then it started decreasing until the 13th month, except for *Bisimwenge*, which showed a slight increase from the 12th month up to the 13th month. For *Nyar-Udota*, there was an increase from the 8th month up to the 9th month, after which the level of hydrogen cyanide started decreasing until the 13th month. By the 13th month, *Bisimwenge* had the highest amount of hydrogen cyanide (181.48 mg/kg), followed by NASE 9 (109.33 mg/kg), TME 14 (105.60 mg/kg), *Nyar-anderiano* (90.00 mg/kg), and finally *Nyar-udota* (88.50 mg/kg) with the lowest amount of the hydrogen cyanide.

This trend could be attributed to the following two opposing factors:

Firstly, cyanogen synthesis, which is based on the expression of the gene CYP79D1/D2, takes place in the young shoots. After the synthesis, it is translocated to the roots [43]. This increases the level of cyanogen in the roots. Secondly, cyanogen re-assimilating based on the expression of the gene β-CAS, where they are exploited for the synthesis of amino acid [12] as well as the enzymes, linamarase and HNL, both of which take part in breaking down the cyanogens upon tissue rapture, based on their expression clustered together. Both cyanogen re-assimilation and the action of linamarase and HNL reduce the level of cyanogen in the roots.

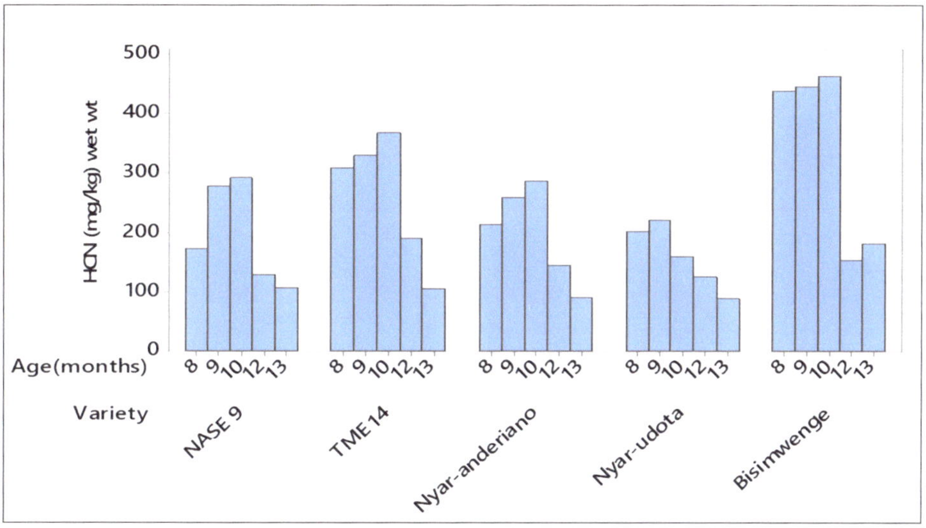

Figure 2.
A graph showing the variation of levels of hydrogen cyanide (HCN) with age in five cassava cultivars (NASE 9, TME 14, 'Nyar-anderiano', 'Nyar-udota', and 'Bisimwenge').

When cyanogen synthesis outweighed, there was an increasing trend of the cyanide level (8–10 months). During the tender age of cassava, cyanogen synthesis is enhanced because of the increased number of young shoots sprouting, which is responsible for the synthesis of the cyanogens. This led to an overall increase in the cyanogen level.

Meanwhile, when cyanogen re-assimilation and action of linamarase and HNL outweighed cyanogen synthesis, there was a decreasing trend in the graph (10–13 months) except for *Nyar-udota* where the decrease was from the 9th month as in **Figure 2**. As the cassava matures, the number of young shoots being produced reduces drastically, leading to a decrease in the amount of cyanogen synthesised, as the rate of cyanogen re-assimilation and action of linamarase and HNL remains fairly constant. Thus, overall, the level of the cyanogens was reduced.

3.2 Hydrogen cyanide levels in the cassava varieties at maturity

The levels of hydrogen cyanide found in the different cassava varieties at maturity (13 months) are presented in **Figure 3**.

The levels of the hydrogen cyanide increased in the order; *Nyar-udota* < *Nyar-anderiano* < NASE 19 < TME 14 < NASE 9 < TME 204 < NASE 3 < NASE 14 < *Terengule* < *Nyar-matia* < *Bisimwenge* < *Nyar-pamitu* < *Nyar-papoga*. In the improved cassava varieties, the HCN level was highest for NASE 14 (116.51 mg/kg) and lowest for NASE 19 (101.84 mg/kg). Amongst the local cassava varieties considered in this study, the cyanide levels in *Nyarudota* (88.5 mg/kg) and *Nyar-anderiano* (90.0 mg/kg) were the lowest, even much lower than for the improved varieties. This was in agreement with what was reported by Afoakwa et al. [44], who generally reported lower HCN in local varieties than in improved ones. The cyanide levels in the other four local cassava varieties were higher than those in the improved varieties (**Figure 3**), contrary to the findings of Afoakwa et al. [44]. Generally, the improved cassava varieties considered in this study have shown

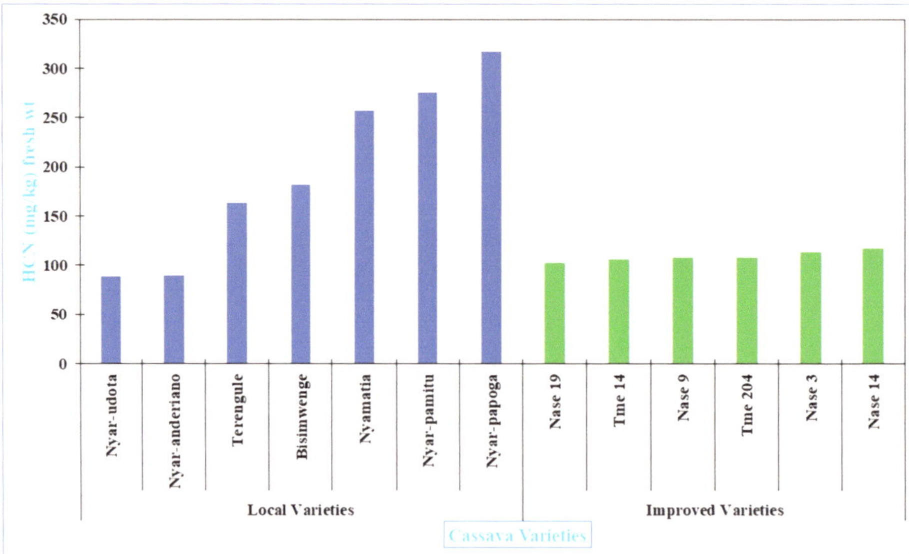

Figure 3.
A graph showing the levels of hydrogen cyanide (in mg/kg) in all the cassava varieties planted at maturity (13 months).

significantly lower levels of hydrogen cyanide (mean value = 108.75) than the local cultivars (mean value = 201.65) ($t = 2.331$, $p = 0.042$). Furthermore, the cyanide level variation in the improved varieties (standard deviation was 5.31) was much lower than in the local cultivars (standard deviation 89.00).

Generally, this trend could be attributed to the fact that in the improved cassava cultivars, the linamarase gene, which is responsible for the disintegration of the cyanogens, has higher transcriptional activity than the bitter cultivars.

It is also possible that there is more inhibition of the cytochrome gene expression that catalyses the first step in the synthesis of linamarin in the improved cassava varieties than the local ones. According to Siritunga and Sayre [43], the linamarin content of cassava roots reduced by 99% in transgenic plants expressing the cytochrome P450 genes (CYP79D1 and CYP79D2) that catalyse the first step in the synthesis of linamarin.

Nonetheless, all the values lie within the cyanide range in cassava root parenchyma of 10–500 mg cyanide equivalents/kg dry weight [43, 45, 46].

3.3 Variation of the level of hydrogen cyanide with fermentation days in

3.3.1 Nyar-papoga cassava variety

The hydrogen cyanide level (mg/kg) dry weight was obtained for cassava samples subjected to varied hours of partial sun-drying and the number of days of fermentation. The trend is presented in the line graph in **Figure 4**.

The level of hydrogen cyanide was high after two (2) days of fermentation but kept reducing steadily until the seventh (7th) day of fermentation. The hydrogen cyanide level in the local cassava variety ('*Nyar-papoga*') varied significantly ($F_{(4, 16)} = 62.48$, $p = 1.49 \times 10^{-9}$) with the number of days of fermentation.

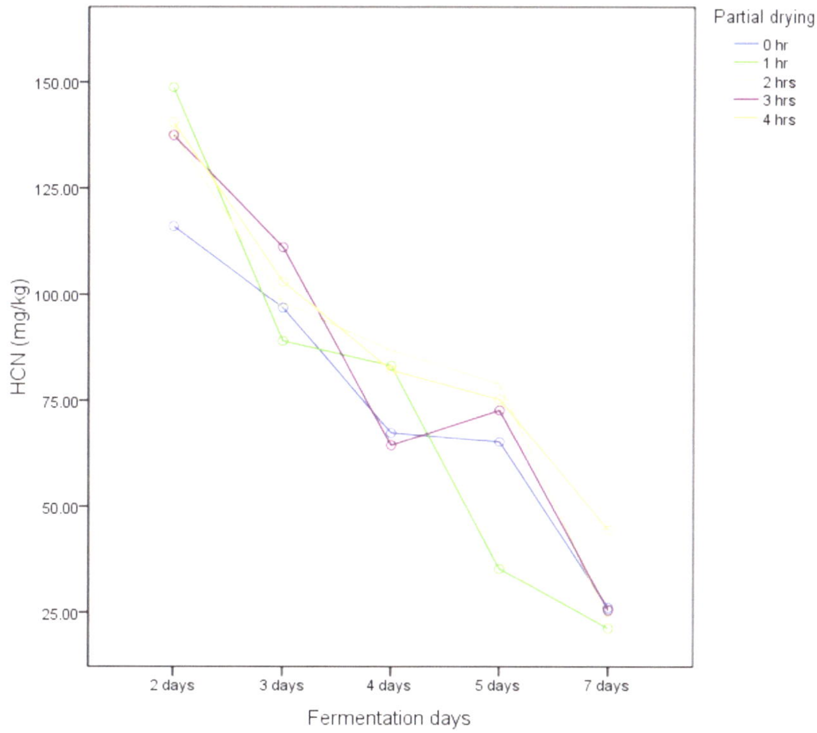

Figure 4.
A graph showing the variation of hydrogen cyanide level in a local cassava variety (Nyar-papoga) with hours of partial drying and fermentation days.

3.3.2 Nyar-udota cassava variety

The hydrogen cyanide level in *Nyar-udota* cassava variety that was subjected to fermentation for a varying number of days was determined. The result is presented in the line graph in **Figure 5**.

The level of hydrogen cyanide in the unfermented (Day 0) dried *Nyar-udota* cassava variety was the highest (52.63 mg/kg). The level of the hydrogen cyanide then decreased steadily with fermentation days until the fourth day. Thereafter, it remained fairly constant until the 10th day of fermentation (18.58 mg/kg), although there was only a slight decrease in the level of hydrogen cyanide. Generally, the hydrogen cyanide level reduced significantly ($F_{(1, 12)} = 19.46$, $p = 8.49 \times 10^{-4}$) with the period of fermentation, with a percentage reduction of about 65% on the 10th day of fermentation.

Fermentation probably causes more cells to rupture, easily bringing about contact between substrate cyanoglycosides and the enzymes, consequently leading to the breakdown of cyanoglycosides to release free HCN [39]. Moreover, heap fermentation generates heat that volatilizes free hydrogen cyanide [39]. However, Lambri et al. [47] and Bradbury [48] revealed that fermentation temperature was not significant because no consistent differences were exhibited between 30 and 35°C fermentation temperatures. However, the warmth generated by fermentation could progressively evaporate the free hydrogen cyanide, which is volatile at 25.7°C [39, 49].

According to Westby [49], the essential features of efficient processing of the cyanogens involve adequate tissue disruption to enable endogenous linamarase to

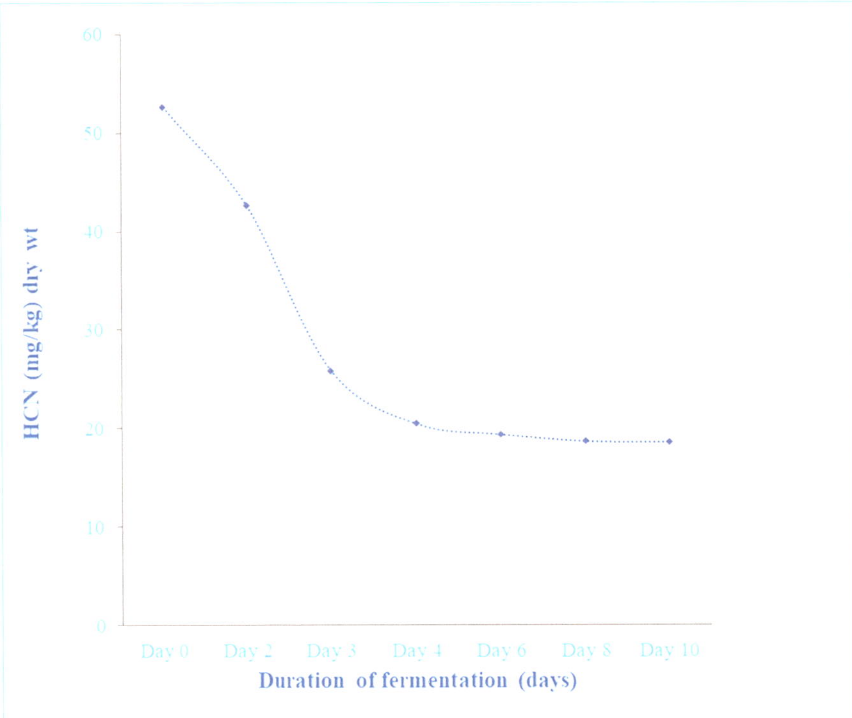

Figure 5.
A graph showing the variation of hydrogen cyanide level with fermentation period in Nyar-udota cassava variety.

come into contact with linamarin and then favourable conditions for the breakdown of acetone cyanohydrin, or conditions that can facilitate spontaneous volatilisation of the compound. In the case of heap-fermented products, microbial growth reduces cyanide content by softening the cassava roots, which increases the contact between endogenous linamarin and linamarase [50].

A series of microorganisms in which the microbial groups, lactic acid bacteria (LAB), and yeasts predominate, characterise natural fermentation [51, 52]. The most frequent LAB species are *Lactobacillus manihotivorans* and *Lactobacillus plantarum* [53]. *L. manihotivorans* exists only during the first period of fermentation, when it may accelerate the rate of degrading starch [52], resulting into contact between linamarase and cyanogenic glycosides, thus, reducing the cyanide level as fermentation progresses. Meanwhile, *L. plantarum*, which is present during all the steps of the fermentative process, acidifies the substrate. Therefore, as the fermentation progresses, there is a gradual decrease in the number of microorganisms due to the increased acidity of the medium [54]. This slows down the fermentation process until it finally stops (**Figure 5**).

4. Conclusions

Improved cassava varieties have lower levels of hydrogen cyanide, and the level does not significantly vary amongst them. The local cassava varieties considered have high levels of hydrogen cyanide except *Nyar-Udota* and *Nyar-anderiano* and there is significant variation of the cyanide levels amongst them. Generally, the improved cassava varieties considered have lower hydrogen cyanide levels than the local ones.

The hydrogen cyanide levels in the cassava cultivars studied were found to be highest at the ages of 8–10 months.

Fermentation reduces the hydrogen cyanide level significantly, and the decrease varies with the fermentation period.

Acknowledgements

The authors are grateful to Mr. Dan Lema of the Government Analytical Laboratory in Kampala, Uganda, for his technical guidance during the analysis of the samples for the cyanide content. The authors also thank the Management of Muni University, who provided the funds used for paying the Open Access publication fees of this Book Chapter.

Conflict of interest

The authors declare no conflict of interest.

Author details

Benson Oloya[1*], Christopher Adaku[2] and Morgan Andama[1]

1 Muni University, Arua, Uganda

2 Mbarara University of Science and Technology, Mbarara, Uganda

*Address all correspondence to: oloyabenson@gmail.com

References

[1] NAADS. Profit Margins on Cassava Production. Kampala; 2020. Available from: https://naads.or.ug/profit-margins-on-cassava-production/

[2] Ano OA, Eke-Okoro NO, Egesi NC. Heavy metals (Cd, Ni and Pb) pollution effects on cassava (Manihot esculenta Crantz). International Journal of Biodiversity Conservation. 2013;**5**:640-646. DOI: 10.5897/IJBC11.029

[3] Bayata A. Review on nutritional value of cassava for use as a staple food. Scientific Journal of Analytical Chemistry. 2019;**7**:83. DOI: 10.11648/j.sjac.20190704.12

[4] Sowcharoensuk C. Cassava industry, Thail. Ind. Outlook. 2018;**20**:1-8

[5] Adebayo K, Abdoulaye S. West Africa Competitiveness Programme Regional Investment Profile – Summary Cassava Value Chain. Geneva; 2022

[6] Bradbury JH, Holloway WD. Chemistry of Tropical Root Crops: Significance for Nutrition and Agriculture in the Pacific. Melbourne: Australian Centre for International Research; 1988

[7] Cardoso AP, Mirione E, Ernesto M, Massaza F, Cliff J, Haque MR, et al. Processing of cassava roots to remove cyanogens. Journal of Food Composition and Analysis. 2005;**18**:451-460. DOI: 10.1016/j.jfca.2004.04.002

[8] Otim-Nape GW, Bua A, Thresh JM, Baguma Y, Ogwal S, Ssemakula G, et al. The Current Pandemic of Cassava Mosaic Virus Disease in East Africa and Its Control. Chatham, UK: Natural Resources Institute (NRI); 2000

[9] O'Brien GM, Wheatley CC, Iglesias C, Poulter N. Evaluation, modification, and comparison of two rapid assays for cyanogens in cassava. Journal of the Science of Food and Agriculture. 1994;**65**:391-399

[10] Grace MR. Elaboration of Cassava. FAO Collection: Plant Protection and Protection. Rome, Italy; 1977

[11] King NLR, Bradbury JH. Bitterness of cassava: Identification of a new apiosyl glucoside and other compounds that affect its bitter taste. Journal of the Science of Food and Agriculture. 1995;**68**:223-230

[12] Jørgensen K, Bak S, Busk PK, Sørensen C, Olsen CE, Puonti-Kaerlas J, et al. Cassava plants with a depleted cyanogenic glucoside content in leaves and tubers. Distribution of cyanogenic glucosides, their site of synthesis and transport, and blockage of the biosynthesis by RNA interference technology. Plant Physiology. 2005;**139**:363-374. DOI: 10.1104/pp.105.065904

[13] Andersen MD, Busk PK, Svendsen I, Møller BL. Cytochromes P-450 from cassava (Manihot esculenta Crantz) Catalyzing the first steps in the biosynthesis of the cyanogenic glucosides Linamarin and Lotaustralin. The Journal of Biological Chemistry. 2000;**275**:1966-1975. DOI: 10.1074/jbc.275.3.1966

[14] McMahon JM, White WLB, Sayre RT. Cyanogenesis in cassava (Manihot esculenta Crantz). Journal of Experimental Botany. 1995;**46**:731-741

[15] Siritunga D, Sayre R. Transgenic approaches for cyanogen reduction in cassava. Journal of AOAC International.

2007;**90**:1451-1455. DOI: 10.1093/jaoac/90.5.1450

[16] Cutler J, Conn E, Grant S. The biosynthesis of cyanogenic glucosides in *Linum usitatissimum* (linen flax) in vitro. Archives of Biochemistry and Biophysics. 1981;**212**:468-474

[17] Yemm RS, Poulton JE. Isolation and characterization of multiple forms of Mandelonitrile from mature black cherry (Prunus serotina Ehrh.) seeds. Archives of Biochemistry and Biophysics. 1986;**247**:440-445

[18] Wajant H, Mundry K. Hydroxynitrile lyase from Sorghum bicolor: A glycoprotein heterodimer. Plant Science. 1993;**89**:127-133

[19] Wajant H, Riedel D, Bent S, MundryK.Immunocytologicallocalization of hydroxynitrile lyases from Sorghum bicolor L. and *Linum usitatissimum* L. Plant Science. 1994;**103**:145-154

[20] Zheng L, Poulton E. Jonathan, temporal and spatial expression of amygdalin hydrolase and (R) - (+)-Mandelonitrile Lyase in black cherry seeds. Plant Physiology. 1995;**109**:31-39

[21] Rosling H. Cassava Cyanide and Epidemic Spastic Paraparesis: A Study in Mozambique on Dietery Cyanide Exposure. Sweden: Uppsala University; 1974

[22] Akintonwa A, Tunwashe O, Onifade A. Fatal and non-fatal acute poisoning attributed to cassava-based meal. Acta Horticulturae. 1994;**375**:285-288

[23] Mlingi N, Poulter NH, Rosling H. An outbreak of acute intoxications from consumption of insufficiently processed cassava in Tanzania. Nutrition Research. 1992;**12**:677-687. DOI: 10.1016/S0271-5317(05)80565-2

[24] Cereda MP, Mattos MCY. Linamarin: The toxic compound of cassava. Journal of Venome Animal Toxins. 1996;**2**. DOI: 10.1590/S0104-79301996000100002

[25] Delange F, Ekpechi LO, Rosling H. Cassava cyanogenesis and iodine deficiency disorders. Acta Horticulturae. 1994;**375**:289-293

[26] Howlett WP, Brubaker GR, Mlingi N, Rosling H. Konzo, an epidemic upper motor neuron disease studied in Tanzania. Brain. 1990;**113**:223-235

[27] Cliff J, Nicala D, Saute F, Givragy R, Azambuja G, Taela A, et al. Konzo associated with war in Mozambique. Tropical Medicine & International Health. 1997;**2**:1068-1074

[28] Nzwalo H, Cliff J. Konzo: From Poverty, Cassava, and Cyanogen Intake to Toxico-Nutritional Neurological Disease. PLOS Neglected Tropical Diseases. 2017;**5**(2011):1-8. DOI: 10.1371/journal.pntd.0001051

[29] Osuntokun BO. Chronic cyanide intoxication of dietary origin and a degenerative neuropathy in Nigerians. Acta Horticulturae. 1994;**375**:311-321

[30] Howlett WP. Konzo; a new human disease entity. Acta Horticulturae. 1994;**375**:323-329

[31] Onabolu AO, Oluwole OSA, Bokanga M, Rosling H. Ecological variation of intake of cassava food and dietary cyanide load in Nigerian communities. Public Health Nutrition. 2001;**4**:871-876

[32] Lang K. Die Rhodanbildung im Tierkörper. Biochemische Zeitschrift. 1933;**259**(1933):243-256

[33] F. Standards. Cyanogenic Glycosides in Cassava and Bamboo Shoots: A Human

Health Risk Assessment. Canberra, Australia/Wellington, New Zealand; 2005

[34] Editor. Bitter cassava to blame for children disability - expert, Dly. Monit 2013. https://www.monitor.co.ug/uganda/news/national/bitter-cassava-to-blame-for-children-disability-expert-1560918

[35] Linley C-K, Chrissie K, Ngoma J, Chipungu F, Mkumbira J, Simukoko S, et al. Bitter cassava and women: An intriguing response to food security. LEISA Mag. 2002;**18**

[36] FAO/WHO. Joint FAO/WHO Food Standards Programme. Rome, Italy; 1991

[37] Bokanga M, Essers S, Poulter N, Rosling H, Tewe O. International workshop on cassava safety (WOCAS). Acta Horticulturae. 1994:375

[38] Akintonwa A, Tunwashe OL. Fatal cyanide poisoning from cassava-based meal. Human & Experimental Toxicology. 1992;**11**:47-49. http://www.ncbi.nlm.nih.gov/pubmed/1354460

[39] Andama M, Lejju JB. Potential of fermentation in detoxifying toxic cassava root tubers. 2012;**2**:1-7

[40] Okigbo NB. Nutritional Implications of Projects Giving High Priority to the Production of Staples of Low Nutritive Quality: The Case of Cassava in the Humid Tropics of West Africa. Tokyo; 1980

[41] Melinfonwu A, James B, Achou K, Weise S, Awah E, Gbaguidi B. Weed Control in Cassava Farms. IPM Field Guide for Extension Agent. Cotonou, Benin: IITA; 2000

[42] FAO. Processing and Utilization of Root and Tuber Crops. Rome, Italy; 2000

[43] Siritunga D, Sayre RT. Generation of cyanogen-free transgenic cassava. Planta. 2003;**217**:367-373

[44] Afoakwa EO, Asiedu C, Budu AS, Chiwona-Karltun L, Nyirendah DB. Chemical composition and cyanogenic potential of traditional and high yielding CMD resistant cassava (Manihot esculenta crantz) varieties. International Food Research Journal. 2012;**19**:175-181

[45] Arguedas P, Cooke RD. Residual cyanide concentration during the extraction of cassava starch. Food Technology. 1982;**17**:251-261

[46] Dufour DL. Cyanide content of cassava (Manihot esculenta, Euphorbiaceae) cultivars used by Tukanoan Indians in Northwest Amazonia. Economic Botany. 1988;**42**:255-266. DOI: 10.1007/BF02858929

[47] Lambri M, Fumi MD, Roda A, De Faveri DM. Improved processing methods to reduce the total cyanide content of cassava roots from Burundi, African. Journal of Biotechnology. 2013;**12**:2685-2691. DOI: 10.5897/AJB2012.2989

[48] Bradbury JH. Simple wetting method to reduce cyanogen content of cassava flour. Journal of Food Composition and Analysis. 2006;**19**:388-393. DOI: 10.1016/j.jfca.2005.04.012

[49] Westby A. Cassava utilization, storage and small-scale processing. In: Cassava Biol. Prod. Util. UK: CABI Publishing; 2001. pp. 281-300. DOI: 10.1079/9780851995243.0281

[50] Essers AJA, Ebong C, van der Grift RM, Nout MJR, Otim-Nape W, Rosling H. Reducing cassava toxicity by heap-fermentation in Uganda. International Journal of Food Sciences

and Nutrition. 1995;**46**:125-136. DOI: 10.3109/09637489509012540

[51] Figueroa C, Davila AM, Pourquié J. Lactic acid bacteria of the sour cassava starch fermentation. Letters in Applied Microbiology. 1995;**21**:126-130. DOI: 10.1111/j.1472-765X.1995.tb01023.x

[52] Ampe F, Sirvent A, Zakhia N. Dynamics of the microbial community responsible for traditional sour cassava starch fermentation studied by denaturing gradient gel electrophoresis and quantitative rRNA hybridization. International Journal of Food Microbiology. 2001;**65**:45-54. DOI: 10.1016/S0168-1605(00)00502-X

[53] Ben Omar N, Ampe F, Raimbault M, Guyot J-P, Tailliez P. Molecular diversity of lactic acid Bacteria from cassava sour starch (Colombia). Systematic and Applied Microbiology. 2000;**23**:285-291. DOI: 10.1016/S0723-2020(00)80016-8

[54] Oyewole O. Characteristics and significance of yeasts' involvement in cassava fermentation for 'fufu' production. International Journal of Food Microbiology. 2001;**65**:213-218. DOI: 10.1016/S0168-1605(01)00431-7